Lecture Notes in Mathematics

Edited by A. Dold and B. Eckmann

408

John Wermer

T0224656

Potential Theory

Second Edition

Springer-Verlag
Berlin Heidelberg New York 1981

Author

John Wermer
Department of Mathematics
Brown University
Providence, RI 02912
U.S.A.

AMS Subject Classifications (1980): 31-02, 31 B XX, 31 B 05, 31 B 10, 31 B 15, 31 B 20

ISBN 3-540-10276-0 Springer-Verlag Berlin Heidelberg New York
ISBN 0-387-10276-0 Springer-Verlag New York Heidelberg Berlin

© by Springer-Verlag Berlin Heidelberg 1981
Printed in Germany

Printing and binding: Beltz Offsetdruck, Hemsbach/Bergstr.
2141/3140-543210

POTENTIAL THEORY

John Wermer

CONTENTS

1. Introduction

Potential theory grew out of mathematical physics, in particular out of the theory of gravitation and the theory of electrostatics. Mathematical physicists such as Poisson and Green introduced some of the central ideas of the subject.

A mathematician with a general knowledge of analysis may find it useful to begin his study of classical potential theory by looking at its physical origins. Sections 2, 5 and 6 of these Notes give in part heuristic arguments based on physical considerations. These heuristic arguments suggest mathematical theorems and provide the mathematician with the problem of finding the proper hypotheses and mathematical proofs.

These Notes are based on a one-semester course given by the author at Brown University in 1971. On the part of the reader, they assume a knowledge of Real Function Theory to the extent of a first year graduate course. In addition some elementary facts regarding harmonic functions are assumed as known. For convenience we have listed these facts in the Appendix. Some notation is also explained there.

Essentially all the proofs we give in the Notes are for Euclidean 3-space R^3 and Newtonian potentials

$$\int \frac{d\mu(y)}{|x-y|} .$$

In Section 4 we discuss the situation in $\mathbb{R}^n$, $n \neq 3$. The modifications needed to go from $\mathbb{R}^3$ to $\mathbb{R}^n$, $n > 3$, are merely technical, so we state most results for $\mathbb{R}^n$, $(n \geq 3)$.

When n = 2, there are significant differences. Potential theory in R^2 is treated in many books concerned with analytic functions. A detailed treatment can be found in the book, "Potential Theory in Modern Function Theory" by M. Tsuji, Tokyo, 1959. The classical treatise on Potential Theory, which carries the subject up to 1930, is O. Kellogg "Foundations of Potential Theory", Springer, 1929.

The present Notes make use of this book and also of the following works: O. Frostman, "Potentiel d'Équilibre et Capacité des Ensembles", Dissertation, Lund, 1935.

Lennart Carleson, "Selected Problems on Exceptional Sets", Van Nostrand Mathematical Studies 13, 1967.

L.L. Helms, "Introduction to Potential Theory", Wiley-Interscience, 1969.

These four books give some historical discussion of Potential Theory and many references to original papers.

We have collected our References at the end of the Notes.

I want to express my thanks to Richard Basener, Yuji Ito and Charles Stanton for their help in preparing these Notes. I am also indebted to Lars Hörmander for valuable conversations. I am grateful to Miss Sandra Spinacci for typing the manuscript.

This work was partially supported by NSF Grant GP-28574.

August 1972 John Wermer

Section 19 on Wiener's Criterion has been added in this second edition. I am very much indebted to David Adams for suggesting that an exposition of Wiener's Criterion fits in naturally with these Notes. The proof given was, apart from some details, shown to me by him.

I also wish to thank Brian Cole who provided Exercise 1.

October 1980 J. W.

2. Electrostatics

We shall consider electric charges distributed in space and the force fields which these charges produce.

Consider an electrically charged body, say negatively charged

If we look at a charged test particle in the presence of this body, we find that a force acts on it. This force depends on the position (x,y,z) of the particle, and on its charge e_1. We have

$$\text{Force on particle} = e_1 \cdot \vec{\mathfrak{E}}(x,y,z).$$

Here $\vec{\mathfrak{E}}(x,y,z)$ is a vector independent of the particle. Thus, associated with our charged body, there is a vector field $\vec{\mathfrak{E}}$ defined everywhere in space outside the body. $\vec{\mathfrak{E}}$ is called the <u>electric field</u> due to the body.

If we have several charged bodies, with electric fields $\vec{\mathfrak{E}}_1, \ldots, \vec{\mathfrak{E}}_n$, then we find that the force on a test particle of charge e_1 now is $e_1 \cdot \sum_{j=1}^{n} \vec{\mathfrak{E}}_j$.

Thus, the electric field due to several bodies is the sum of their individual fields.

If the body is idealized to be a point charge e located at a point A then the electric field at each point B has direction AB and magnitude $\frac{e}{r^2}$ (Coulomb's Law), where r = distance (A,B), if $e > 0$.

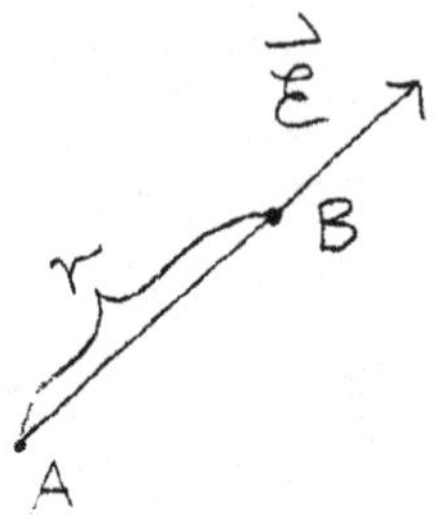

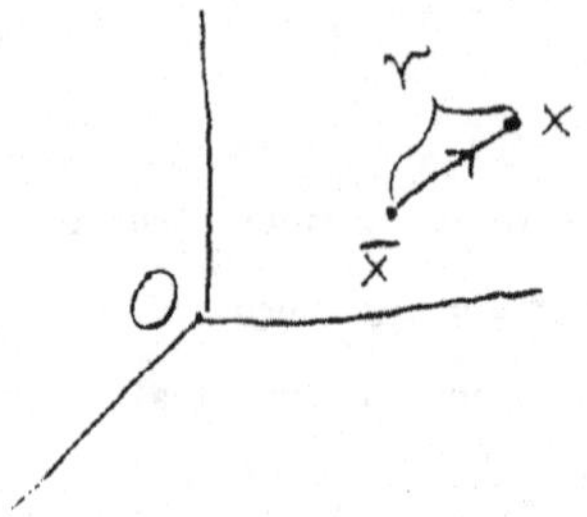

In vector notation, a point charge e at $\bar{x}$ induces a field $\vec{\mathfrak{E}}$ such that

(2.1) $$\vec{\mathfrak{E}}(x) = \frac{e}{r^2} \cdot \frac{x - \bar{x}}{r} = \frac{e}{r^3}(x-\bar{x})$$

for all $x \in \mathbb{R}^3$, where $r = |x-\bar{x}|$. Note that $\vec{\mathfrak{E}}$ is singular at $\bar{x}$.

At the end of the 18th century it was observed by Lagrange that $\exists$ a scalar function φ with

$$\vec{\mathfrak{E}} = -\text{grad}\,\varphi,$$

where $\varphi(x) = \dfrac{e}{|x-\bar{x}|}$. (Verify)

In general, a force field $\vec{\mathfrak{F}}$ is said to have a potential function U if

$$\vec{\mathfrak{F}} = -\text{grad}\,U, \text{ i.e.,}$$

$$\vec{\mathfrak{F}} = (F_1,F_2,F_3),\ F_1 = -U_x,\ F_2 = -U_y,\ F_3 = -U_z.$$

(The minus sign is convention.) Some force fields have potential functions and others don't.

Consider a force field $\vec{\mathfrak{F}}$ that does: $\vec{\mathfrak{F}} = -\text{grad}\,U$. Fix two points P,Q and a path β: $x = x(s)$, $y = y(s)$, $z = z(s)$, from P to Q, where the parameter s is arc-length.

The work W done by the field $\vec{\mathfrak{F}}$ in moving a particle from P to Q is defined by

$$W = \int_P^Q F_t ds,$$

where F_t = tangential component of $\mathfrak{F}$.

The unit tangent vector to β is

$$\vec{T} = \left(\frac{dx}{ds}, \frac{dy}{ds}, \frac{dz}{ds}\right).$$

Hence

$$\begin{aligned} F_t &= F_1 \frac{dx}{ds} + F_2 \frac{dy}{ds} + F_3 \frac{dz}{ds} \\ &= -U_x \frac{dx}{ds} - U_y \frac{dy}{ds} - U_z \frac{dz}{ds} \\ &= -\frac{d}{ds} U(x(s),y(s),z(s)). \end{aligned}$$

Hence

$$W = -\int_P^Q \left(\frac{d}{ds} U\right) ds = -(U(Q) - U(P)).$$

Thus: the work done by the field in moving the particle from P to Q = -change in potential functions from P to Q.

Suppose now we are given n point charges e_i where e_i is located at x_i, $i = 1,2,\ldots n$. What is the field $\vec{\mathfrak{E}}$ they produce? Using (2.1), we get

$$\vec{\mathfrak{E}} = \sum_i \frac{e_i}{|x-x_i|^3}(x-x_i) = -\text{grad}\,\varphi,$$

(2.2) $$\varphi(x) = \sum_{i=1}^n \frac{e_i}{|x-x_i|}.$$

Suppose now we have charges continuously distributed over a body, rather than a finite set of point charges. That is, suppose there is a continuous function ρ defined on the body Ω such that for every portion B of Ω the charge in B is given by

$$\int_B \rho dV,$$

where dV is the volume element. We call ρ the <u>charge density</u>. How to find $\vec{\mathfrak{E}}$ now?

We divide the body Ω up into little bits B_i, $i = 1,\dots n$ and in each bit choose a point x_i. Then the charge in B_i is approximately $= \rho(x_i)\,\Delta V_i$. Replace the charge in B_i by a point charge $\rho(x_i)\,\Delta V_i$ at x_i. The field due to this finite system of point charges $\vec{\mathfrak{E}}^n$ is given by

$$\vec{\mathfrak{E}}^n(x) = \sum_{i=1}^{n} \frac{\rho(x_i)\,\Delta V_i}{|x-x_i|^3}(x-x_i).$$

Letting $n \to \infty$, we get, for $x \in \mathbb{R}^3\setminus\Omega$:

$$\vec{\mathfrak{E}}(x) = \int_\Omega \frac{1}{|x-\xi|^3}(x-\xi)\rho(\xi)dV. \tag{2.3}$$

Putting

$$\varphi(x) = \int_\Omega \frac{\rho(\xi)dV}{|x-\xi|}, \quad x \in \mathbb{R}^3\setminus\Omega, \tag{2.4}$$

we get as before that $\vec{\mathfrak{E}} = -\text{grad}\,\varphi$.

The formulas (2.2) and (2.4) for the potential function φ suggest the following generalization. Let μ be a positive Borel measure on $\mathbb{R}^3$ with finite total mass, i.e.,

$$\int_{\mathbb{R}^3} d\mu(y) < \infty.$$

Definition 2.1: The potential of μ is the function U^μ defined by

$$U^\mu(x) = \int_{\mathbb{R}^3} \frac{d\mu(y)}{|x-y|} . \tag{2.5}$$

We regard $U^\mu(x)$ as defined for all x in $\mathbb{R}^3$, with values possibly $= \infty$ at certain x. For x outside the support of μ, $U^\mu(x) < \infty$.

Exercise 2.1: Assume μ has compact support. Fix R.

(a) Show $\int_{|x|<R} U^\mu(x)dx < \infty$

(b) Deduce $U^\mu(x) < \infty$ a.e. $-dx$.

Note that formulas (2.2) and (2.4) are special cases of (2.5), with μ a finite sum of point masses, resp. $\mu = \rho(x)dx$. We can get a new example by taking $\mu = dx_1dx_2$ on a disk D in the x_1x_2-plane. Then

$$U^\mu(x) = \iint_D \frac{d\xi_1 d\xi_2}{|x-\xi|} .$$

Exercise 2.2: Assume μ has compact support K. Show that, in $\mathbb{R}^3\setminus K$, U^μ is in C^∞ and satisfies $\Delta U^\mu = 0$, i.e. U^μ is harmonic.

We should like to know: if μ is a measure and we know its potential U^μ, how do we find μ? I.e., how can we recover μ from U^μ?

It turns out that this problem is connected with some physical considerations regarding the "flux" of an electric field. The notion of flux is attached to a vector-field.

Let $\vec{v}$ be a vector-field in $\mathbb{R}^3$. We regard it as the velocity field of a steady, incompressible fluid flow. Given a piece of surface σ, it makes sense to speak about the quantity of "fluid" passing through σ in unit time. This quantity is called the flux through σ.

If σ is a small (oriented) planar piece, $\vec{v}$ is essentially constant near σ, and the fluid passing through σ in unit time is contained in the parallelopiped

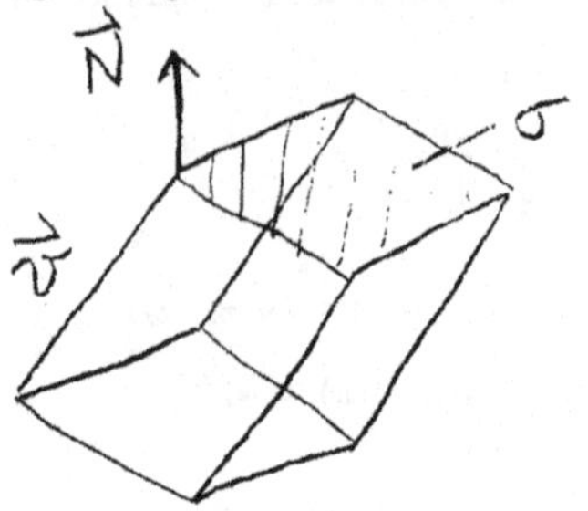

where $\vec{N}$ is the unit normal to σ. Thus: flux through σ = volume of this solid = $(\vec{v}\cdot\vec{N})$ · area of σ.

Approximating an arbitrary oriented smooth surface Σ by polygonal surfaces, and calculating the flux across the polygonal surfaces by means of the preceding formula, we obtain:

$$\text{flux through } \Sigma = \int_{\Sigma} \vec{v}\cdot\vec{N}dS,$$

where $\vec{N}$ is the normal to Σ and dS is the element of area.

The Divergence Theorem states, in particular, that if Ω is a region in $\mathbb{R}^3$ and $\vec{v}$ a vector field defined and smooth on $\overline{\Omega}$, then

$$(2.6) \qquad \int_{\partial\Omega} \vec{v}\cdot\vec{N}dS = \int_{\Omega} \operatorname{div} \vec{v}\cdot dV,$$

where $\vec{N}$ is the outer normal to $\partial\Omega$, dV is the element of volume,

$$v = (v_1, v_2, v_3), \ \operatorname{div} \vec{v} = \sum_{i=1}^{3} \frac{\partial v_i}{\partial x_i}.$$

Let now $\vec{\mathfrak{E}}$ be the electric field due to a point charge e at O. We can speak of the flux of $\vec{\mathfrak{E}}$ across a surface, since $\vec{\mathfrak{E}}$ is a vector field, and we ignore the question of what "fluid" is flowing here.

We claim

(2.7) $$\operatorname{div} \vec{\mathfrak{E}} = 0 \quad \text{for} \quad x \neq 0.$$

For, with $r = \left(\sum_i x_i^2 \right)^{1/2}$, $\vec{\mathfrak{E}}(x) = e\left(\frac{x_1}{r^3}, \frac{x_2}{r^3}, \frac{x_3}{r^3}\right)$, so $\operatorname{div} \vec{\mathfrak{E}} = e \sum_i \frac{\partial}{\partial x_i}\left(\frac{x_i}{r^3}\right)$.

Noting

(2.8) $$\frac{\partial r}{\partial x_j} = \frac{x_j}{r} \quad \text{for all} \quad j,$$

$$\operatorname{div} \vec{\mathfrak{E}} = e \sum_i \left(\frac{r^3 - x_i 3r^2 \cdot \frac{x_i}{r}}{r^6} \right)$$

$$= \frac{e}{r^6} \left\{ 3r^3 - 3r \sum_i x_i^2 \right\} = 0,$$

as claimed.

Let Σ be a closed surface in $\mathbb{R}^3$.

<u>Result 1</u>: If Σ surrounds the charge, i.e. if 0 lies inside Σ, then the flux of $\vec{\mathfrak{E}}$ across $\Sigma = 4\pi e$.

For choose a ball: $|x| \leq \epsilon$, inside Σ. Let Ω_ϵ be the region between $|x| = \epsilon$ and Σ. By (2.6)

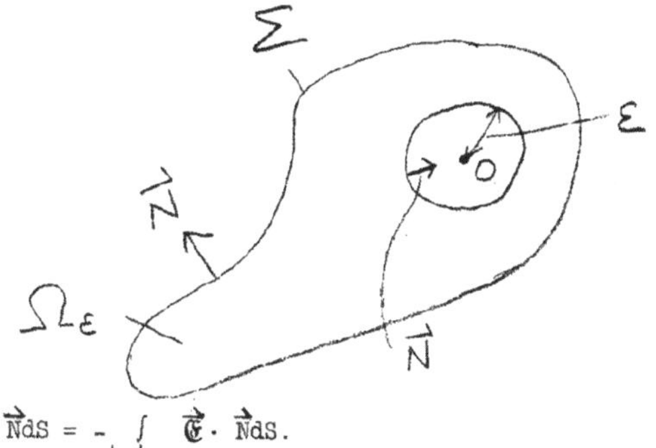

$$\int_{\partial\Omega_\epsilon} \vec{\mathfrak{E}} \cdot \vec{N} dS = \int_{\Omega_\epsilon} \operatorname{div} \vec{\mathfrak{E}}\, dS,$$

and the right side $= 0$ by (2.7).

Since $\partial\Omega_\epsilon = \Sigma \cup \{|x| = \epsilon\}$,

(2.9) $$\int_{\Sigma} \vec{\mathfrak{E}} \cdot \vec{N} dS = - \int_{|x|=\epsilon} \vec{\mathfrak{E}} \cdot \vec{N} dS.$$

On $|x| = \epsilon$, $\vec{\mathfrak{E}} = e \cdot \frac{x}{\epsilon^3}$, $\vec{N} = -\frac{x}{\epsilon}$. So the right side in (2.9) is

$$- \int_{|x|=\epsilon} e \cdot - \frac{\epsilon^2}{\epsilon^4} \, dS = 4\pi e,$$

and so (2.9) gives Result 1.

Result 2: If 0 lies outside Σ, the flux through $\Sigma = 0$.

The proof is just as above, only briefer.

Next, let $\vec{\mathfrak{E}}$ be the field produced by a finite set of point charges, e_i at x_i, $i = 1,2,\ldots,n$. Let $x_{i_1}, x_{i_2}, \ldots, x_{i_k}$ be a subset of these points and Σ a surface containing these points inside it, while the other charges lie outside Σ. Letting Ω_ϵ denote the region bounded by Σ and the spheres $|x - x_{i_\nu}| = \epsilon$, $\nu = 1,\ldots,k$, using the fact that $\operatorname{div} \vec{\mathfrak{E}} = 0$ in Ω_ϵ and using (2.6), we get

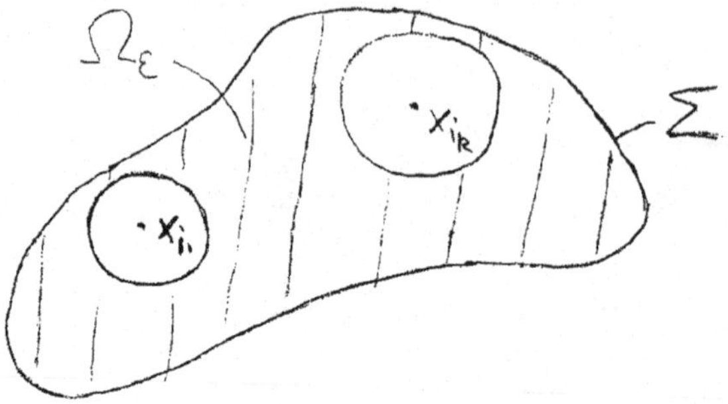

$$\int_\Sigma \vec{\mathfrak{E}} \cdot \vec{N} dS = \sum_{\nu=1}^{k} - \int_{|x-x_{i_\nu}|} \vec{\mathfrak{E}} \cdot \vec{N} dS.$$

Each term of the sum is computed, as earlier, to $= 4\pi e_{i_\nu}$. Hence

$$\int_\Sigma \vec{\mathfrak{E}} \cdot \vec{N} dS = \sum_{\nu=1}^{k} 4\pi e_{i_\nu} .$$

But $\sum_{\nu=1}^{k} e_{i_\nu}$ is the total charge contained inside Σ. So we have:

Result 3: The flux through Σ equals $4\pi \cdot$ the total charge contained inside Σ.

Consider now a continuous distribution of charge on a body Ω, with charge density

ρ. Let Σ be a surface bounding a portion B of Ω.

As before, replace the actual charge in Ω by point charges. Let $x_1, \ldots, x_n$ be the location of the point charges in B, where $\rho(x_i) \Delta V_i$ is the amount of charge at x_i and ΔV_i is the volume of the corresponding portion of B.

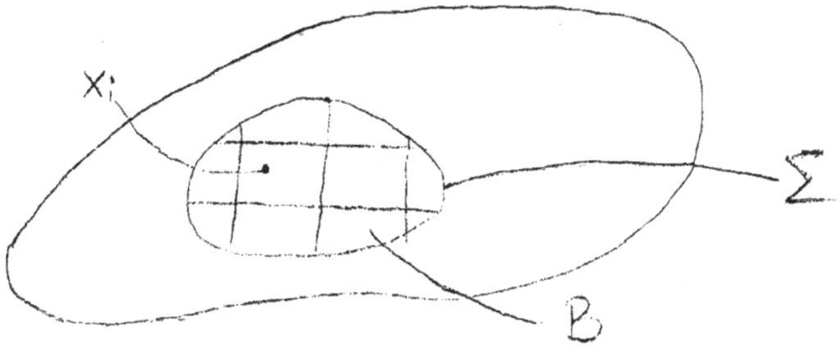

The flux through Σ due to all point charges in Ω is then

$$4\pi \sum_i \rho(x_i) \Delta V_i$$

by Result 3. As $n \to \infty$ we obtain:

$$\int_\Sigma \vec{\mathfrak{E}} \cdot \vec{N} dS = 4\pi \int_B \rho dV,$$

where $\vec{\mathfrak{E}}$ is the field due to the charge distribution on Ω.

Assuming now that the components of $\vec{\mathfrak{E}}$ are smooth on B (a shaky assumption since we are now looking at the field in places where there is charge), we get from (2.6) that

(2.10) $$\int_B \operatorname{div} \vec{\mathfrak{E}} dV = 4\pi \int_B \rho dV.$$

Since B was an arbitrary sub-region of Ω, we conclude, by letting B shrink to a point x_0 in Ω, that

(2.11a) $$\operatorname{div} \vec{\mathfrak{E}}(x_0) = 4\pi\rho(x_0).$$

An exactly parallel argument yields for a point x_o outside Ω

$$\text{div } \vec{\mathfrak{E}}(x_o) = 0. \tag{2.11b}$$

We can go one step further. By (2.4), $\vec{\mathfrak{E}} = -\text{grad } \varphi$, where

$$\varphi(x) = \int_\Omega \frac{\rho(\xi)dV}{|x-\xi|},$$

for x outside Ω. Suppose this relation also holds in Ω. Then, using the fact that div grad $\varphi = \Delta\varphi$, where Δ = Laplacian = $\sum_{i=1}^{3} \frac{\partial^2}{\partial x_i^2}$, we have

$$-\Delta\varphi = 4\pi\rho \quad \text{on } \Omega. \tag{2.12}$$

Note 1: From a mathematical point of view we clearly have not proved (2.12). However, we have a reasonable conjecture: Under suitable hypotheses on ρ, the function $\varphi(x) = \int \frac{\rho(\xi)dV}{|x-\xi|}$ is a solution of equation (2.12).

We shall prove this conjecture in the next section.

Note 2: Equation (2.11a) is one of Maxwell's equations for the electro-magnetic field.

3. Poisson's Equation

We begin with some Lemmas regarding the smoothness of the potential of a measure on the support of the measure.

For $\mu = \rho dx$, where ρ is a summable function and dx is Lebesgue measure in $\mathbb{R}^3$, we write U^ρ for U^μ.

From Exercise 2.1 we know that U^ρ is finite a.e. What more can we say?

Lemma 3.1: *Let ρ be a bounded and measureable function of compact support in $\mathbb{R}^3$, $\rho \geq 0$. Then U^ρ is continuous at each point of $\mathbb{R}^3$.*

Proof: Given $F(x,y)$ continuous in $\mathbb{R}^3 \times \mathbb{R}^3$ and put

$$\Phi(x) = \int F(x,y)\rho(y)dy, \qquad x \in \mathbb{R}^3.$$

Choose R with $\operatorname{supp} \rho \subseteq \{|y| < R\}$. Given ϵ, choose δ such that if x_1, x_2 lie in $|x| < R$ and if $|x_1 - x_2| < \delta$, then $|F(x_1,y) - F(x_2,y)| < \epsilon$ for all y in $|y| < R$. Then

$$|\Phi(x_1) - \Phi(x_2)| \leq \int |F(x_1,y) - F(x_2,y)|\rho(y)dy < \epsilon \int \rho(y)dy.$$

Hence Φ is continuous everywhere.

$$U^\rho(x) = \int \frac{\rho(\xi)d\xi}{|x-\xi|}.$$

The preceding does not yield that U^ρ is continuous, since $F(x,y) = \dfrac{1}{|x-y|}$ fails to be continuous. However, we can approximate $\dfrac{1}{|x-y|}$ by a continuous kernel as follows: Put

$$K_n(t) = \begin{cases} \frac{1}{t} & t > \frac{1}{n} \\ n & 0 \leq t < \frac{1}{n} \end{cases} .$$

Put $U_n(x) = \int K_n(|x-\xi|)\rho(\xi)d\xi$. For each n, U_n is then continuous by the above. Also for $n > m$:

$$0 \leq K_n(t) - K_m(t) = \begin{cases} 0 & t \geq \frac{1}{m} \\ \leq \frac{1}{t} & t < \frac{1}{m} \end{cases}$$

$$\begin{aligned} U_n(x) - U_m(x) &= \int (K_n(|x-\xi|) - K_m(|x-\xi|))\rho(\xi)d\xi \\ &= \int_{|x-\xi|<1/m} (K_n(|x-\xi|) - K_m(|x-\xi|))\rho(\xi)d\xi \\ &\leq \int_{|x-\xi|<1/m} \frac{\rho(\xi)}{|x-\xi|} d\xi \leq C \int_{|x-\xi|<1/m} \frac{d\xi}{|x-\xi|} , \end{aligned}$$

since ρ is bounded, $\leq C \int_{|\xi'|<1/m} \frac{d\xi'}{|\xi'|}$, which $\to 0$ as $m \to \infty$ as we see by writing the integral in spherical coordinates.

Thus $\{U_n\}$ is uniformly Cauchy in $\mathbb{R}^3$. Since $U^\rho(x) = \lim_{n \to \infty} U_n(x)$ for each x, by the monotone convergence theorem, U^ρ is continuous in $\mathbb{R}^3$. q.e.d.

Lemma 3.2: Let $\rho \in C_0^k(\mathbb{R}^3)$. Then $U^\rho \in C^k(\mathbb{R}^3)$.

Note: We shall use, without special reference, the result that if σ is a finite measure of compact support and $K(x,y)$ a function continuous in x and y and differentiable in x, then

$$\frac{\partial}{\partial x_j}\{\int K(x,y)d\sigma(y)\} = \int \frac{\partial K}{\partial x_j}(x,y)d\sigma(y)$$

Proof of Lemma: Choose R with $\operatorname{supp}\rho \subset \{|x| < R\}$. Fix x, $|x| < R$.

$$U^\rho(x) = \int_{|\zeta|<R} \frac{\rho(\zeta)dV}{|x-\zeta|} = \int_{|x+\eta|<R} \frac{\rho(x+\eta)}{|\eta|}dV = \int_{|\eta|<2R} \rho(x+y)\frac{dV}{|\eta|}.$$

Since

$$\int_{|\eta|<2R} \frac{dV}{|\eta|} < \infty$$

and

$$\rho(x+\eta) \in C^k,$$

$$\frac{\partial U^\rho}{\partial x_j}(x) = \int_{|\eta|<2R} \frac{\partial\rho}{\partial x_j}(x+\eta)\frac{dV}{|\eta|},$$

which is continuous for all x. Since $\rho \in C_o^k$, it follows by repeated differentiation that $U^\rho \in C^k$. q.e.d.

Theorem 3.3: Let $\varphi \in C_o^2(\mathbb{R}^3)$. Then

$$\varphi(x) = -\frac{1}{4\pi}\int \frac{\Delta\varphi(\zeta)d\zeta}{|x-\zeta|}, \qquad x \in \mathbb{R}^3. \tag{3.1}$$

Proof: Fix $x \in \mathbb{R}^3$. Choose R such that $|x| < R$ and $\varphi(\zeta) = 0$ for $|\zeta| > R - \delta$ where $\delta > 0$. Fix $\epsilon > 0$ and let Ω_ϵ be the region shown

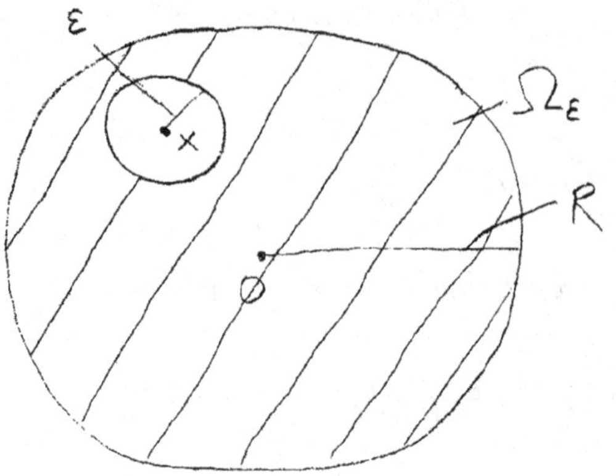

Note that $u(\zeta) = \frac{1}{|x-\zeta|}$ is harmonic in Ω_ϵ and apply G4 (see Appendix) to u and φ in Ω_ϵ. Then

(3.2) $$\int_{\partial\Omega_\epsilon} \left\{ \frac{1}{|x-\zeta|} \frac{\partial\varphi}{\partial n} - \varphi \frac{\partial}{\partial n} \left(\frac{1}{|x-y|} \right) \right\} dS = \int_{\Omega_\epsilon} \left\{ \frac{1}{|x-\zeta|} \Delta\varphi(\zeta) \right\} dV.$$

Introduce spherical coordinates with pole at x. Then

$$-\frac{\partial}{\partial n}\left(\frac{1}{|x-\zeta|}\right) = \frac{\partial}{\partial r}\left(\frac{1}{r}\right) = -\frac{1}{r^2}.$$

Hence

$$\int_{|x-\zeta|=\epsilon} \left\{ \frac{1}{|x-\zeta|} \frac{\partial\varphi}{\partial n} - \varphi \frac{\partial}{\partial n} \left(\frac{1}{|x-\zeta|} \right) \right\} dS = \int_{r=\epsilon} \frac{1}{\epsilon} \frac{\partial\varphi}{\partial n} dS - \int_{r=\epsilon} \varphi(\zeta) \frac{1}{\epsilon^2} dS.$$

Since $\frac{\partial\varphi}{\partial n} = O(1)$ as $\epsilon \to 0$ and $\varphi(\zeta) = \varphi(x) + o(1)$ on $r = \epsilon$, the integral =

$$\frac{1}{\epsilon} O(\epsilon^2) - \frac{1}{\epsilon^2} (\varphi(x) + o(1)) 4\pi\epsilon^2 = o(1) - 4\pi\varphi(x).$$

(3.2) thus gives, using $\varphi = \frac{\partial\varphi}{\partial n} = 0$ on $|\zeta| = R$,

$$o(1) - 4\pi\varphi(x) = \int_{\Omega_\epsilon} \frac{\Delta\varphi(\zeta)}{|x-\zeta|} dV.$$

Letting $\epsilon \to 0$, we arrive at (3.1).

The following theorem gives a partial proof of the conjecture regarding the equation (2.12) above.

__Theorem 3.4__: __Let__ $\rho \in C_0^2(\mathbb{R}^3)$. __Then__

$$\Delta U^\rho = -4\pi\rho.$$

__Proof__: By Lemma 3.2, $U^\rho \in C^2$. Fix $f \in C_0^2$. By G4, we have for large R

$$\int_{|x|<R} U^\rho \, \Delta f dx = \int_{|x|<R} f \, \Delta U^\rho dx. \tag{3.3}$$

The left side =

$$\int_{|x|<R} \left\{ \int_{|\zeta|<R} \frac{\rho(\zeta)d\zeta}{|x-\zeta|} \right\} \Delta f(x)dx.$$

By Fubini's theorem, this =

$$\int_{|\zeta|<R} \rho(\zeta) \left\{ \int_{|x|<R} \frac{\Delta f(x)}{|x-\zeta|} dx \right\} d\zeta.$$

By Theorem 3.3 this in turn =

$$\int_{|\zeta|<R} \rho(\zeta)(-4\pi f(\zeta))d\zeta.$$

Hence (3.3) gives

$$\int_{|x|<R} f \, \Delta U^\rho dx = -4\pi \int_{|\zeta|<R} f\rho d\zeta.$$

Since f was arbitrary, we conclude

$$\Delta U^\rho = -4\pi\rho. \qquad \text{q.e.d.}$$

What if ρ is smooth in a region but does not vanish on the boundary?

Theorem 3.5: Let Ω be a bounded region in $\mathbb{R}^3$ and let $\rho \in C^2(\Omega_1)$ where Ω_1 is a region containing $\bar{\Omega}$. Then $U^\rho \in C^2(\Omega)$ and $\Delta U^\rho = -4\pi\rho$ in Ω.

Proof: Extend ρ to a function $\bar{\rho}$ with $\bar{\rho} \in C_o^2$, $\bar{\rho} = \rho$ in Ω, supp $\bar{\rho} \subset \Omega_1$.

$$(3.4) \qquad \int_{\mathbb{R}^3} \frac{\bar{\rho}(\zeta)d\zeta}{|x-\zeta|} = \int_{\Omega} \frac{\bar{\rho}d\zeta}{|x-\zeta|} + \int_{\Omega_1\setminus\Omega} \frac{\bar{\rho}d\zeta}{|x-\zeta|}$$

The left side $\in C^2(\mathbb{R}^3)$ and by the last theorem has as its Laplacian $-4\pi\bar{\rho}$. Taking Laplacians on both side in (3.4) and noting that the second integral on the right in (3.4) is harmonic in Ω, we get

$$-4\pi\bar{\rho} = \Delta\left\{\int_{\Omega} \frac{\bar{\rho}d\zeta}{|x-\zeta|}\right\}$$

in Ω. Noting that $\bar{\rho} = \rho$ in Ω, this gives that $-4\pi\rho = \Delta U^\rho$ in Ω, as claimed. (3.4) also yields that $U^\rho \in C^2(\Omega)$.

Note 1: If ρ is smooth in a bounded region Ω, up to $\partial\Omega$, and if

$$U^\rho(x) = \int_{\Omega} \frac{\rho(\zeta)d\zeta}{|x-\zeta|},$$

then U^ρ will, in general, have second partial derivatives which are discontinuous on $\partial\Omega$. For $\Delta U^\rho = 0$ outside $\bar{\Omega}$ and $\Delta U^\rho = -4\pi\rho$ in Ω, so unless $\rho \equiv 0$ on $\partial\Omega$, discontinuity must occur.

Note 2: The equation

$$\Delta\varphi = F,$$

which we have been studying, was introduced by the French mathematical physicist Siméon-Denis Poisson (1781-1840) for the gravitational potential inside an attracting body.

4. Fundamental Solutions

There is a trick by which we can "differentiate" functions which don't have derivatives in the usual sense. This trick is the theory of distributions of Laurent Schwartz.

In this theory functions are regarded as linear functionals on a certain vector space, and the distribution "derivative" of a function is another such linear functional This point of view will turn out to be natural and helpful when studying potentials of measures.

Fix $n \geq 1$ and put $C_0^\infty = C_0^\infty(\mathbb{R}^n)$.

Definition 4.1: A distribution on $\mathbb{R}^n$ is a linear map $\Phi\colon C_0^\infty \to \mathbb{R}$ which is continuous in the following sense. If K is a compact set in $\mathbb{R}^n$ and if $\{u_j\}$ is a sequence of elements of C_0^∞ with $\operatorname{supp} u_j \subset K$ for each j and if u_j and each of its derivatives $\to 0$ uniformly, then $\Phi(u_j) \to 0$.

Definition 4.2: L^1_{loc} is the space of all functions u locally summable on $\mathbb{R}^n$, i.e., such that u is measurable and

$$\int_B |u| dx < \infty$$

for every ball B.

Definition 4.3: If $u \in L^1_{loc}$, $\tilde{u}$ is the distribution defined by

$$\tilde{u}(v) = \int uv dx, \quad \text{all } v \in C_0^\infty .$$

Definition 4.4: If Φ is a distribution on $\mathbb{R}^n$, and j is given, $\frac{\partial \Phi}{\partial x_j}$ is the distribution defined by

$$\frac{\partial \Phi}{\partial x_j}(v) = -\Phi\left(\frac{\partial v}{\partial x_j}\right)$$

for all $v \in C_0^\infty$.

We are now ready to "differentiate" an arbitrary function u in L^1_{loc}: the "derivative" $\frac{\partial}{\partial x_j}$ of u is the distribution $\frac{\partial \tilde{u}}{\partial x_j}$. Since $\frac{\partial \tilde{u}}{\partial x_j}$ is a distribution, $\frac{\partial}{\partial x_k}\left(\frac{\partial \tilde{u}}{\partial x_j}\right)$ is again defined as a distribution, so u now has "derivatives" of all orders. Let $u \in C^1(\mathbb{R}^n)$. Then the distribution derivative of u (defined above) coincides with the ordinary derivative. Precisely, for each j

(4.1) $$\frac{\partial \tilde{u}}{\partial x_j} = \widetilde{\left(\frac{\partial u}{\partial x_j}\right)}.$$

Proof: $\frac{\partial \tilde{u}}{\partial x_j}(v) = -\int u \frac{\partial v}{\partial x_j} dx$ for $v \in C_0^\infty$. It follows from the divergence theorem that for $v \in C_0^\infty$,

$$\int \frac{\partial u}{\partial x_j} v dx = -\int u \frac{\partial v}{\partial x_j} dx.$$

Thus $\frac{\partial \tilde{u}}{\partial x_j}(v) = \widetilde{\frac{\partial u}{\partial x_j}}(v)$. So (4.1) holds.

Repeated application of (4.1) yields that if $u \in C^k(\mathbb{R}^n)$ and if L is a linear differential operator with constant coefficients whose order $\leq k$, then

(4.2) $$L(\tilde{u}) = (\widetilde{Lu}).$$

By Exercise 2.1, if μ is a measure of compact support on $\mathbb{R}^3$, then $U^\mu \in L^1_{loc}$. Hence with $\Delta = \sum_{i=1}^{3} \frac{\partial^2}{\partial x_i^2}$, $\Delta(U^\mu)$ is defined as distribution.

Theorem 4.1: Let μ be a measure of compact support on $\mathbb{R}^3$. Then

(4.3) $$\Delta U^\mu = -4\pi\mu.$$

Note: We intepret the right-hand side in (4.3) as the distribution

$$v \to \int -4\pi v d\mu, \qquad v \in C_0^\infty .$$

Also, by abuse of notation we write ΔU^μ for $\Delta\widetilde{U^\mu}$.

<u>Proof</u>: It is easily checked that for $u \in L^1_{loc}$, j fixed,

$$\frac{\partial^2 \tilde{u}}{\partial x_j^2}(v) = \int u \frac{\partial^2 v}{\partial x_j^2} dx, \qquad v \in C_0^\infty ,$$

whence

$$\Delta\tilde{u}(v) = \int u \Delta v dx.$$

Applying this with $u = U^\mu$ we get for $f \in C_o^\infty$:

$$\begin{aligned}\Delta U^\mu(f) &= \int U^\mu(x)\Delta f(x)dx \\ &= \int \left\{ \int \frac{d\mu(\zeta)}{|x-\zeta|} \right\} \Delta f(x)dx = \int \left\{ \int \frac{\Delta f(x)dx}{|x-\zeta|} \right\} d\mu(\zeta) \\ &= -4\pi \int f(\zeta)d\mu(\zeta),\end{aligned}$$

by (3.1). (4.3) thus holds.

<u>Corollary 1</u>: <u>If</u> μ_1,μ_2 <u>are measures of compact support on</u> $\mathbb{R}^3$ <u>with</u> $U^{\mu_1} = U^{\mu_2}$, a.e. -dx, <u>then</u> $\mu_1 = \mu_2$.

<u>Proof</u>: Since $U^{\mu_1} = U^{\mu_2}$, a.e., the two functions in L^1_{loc} determines the same distribution, and so by Theorem 4.1, $\mu_1 = \mu_2$.

<u>Corollary 2</u>: $\frac{1}{|x|} \in L^1_{loc}$.

(4.4) $$\Delta\left(\frac{1}{|x|}\right) = -4\pi\delta,$$

where δ is the distribution on $\mathbb{R}^3$ with $\delta(v) = v(0)$, all v.

Proof: Let μ_0 be the unit point mass at 0

$$\Delta U^{\mu_0} = -4\pi\mu_0$$

by Theorem 4.1. But $U^{\mu_0}(x) = \int \frac{d\mu_0(\zeta)}{|x-\zeta|} = \frac{1}{|x|}$ and the measure μ_0 induces the distribution δ. q.e.d.

We now look at the corresponding situation in $\mathbb{R}^n$, n arbitrary. Denote by δ the distribution on $\mathbb{R}^n$ given by

$$\delta(v) = v(0), \qquad \text{all } v \in C_0^\infty(\mathbb{R}^n).$$

Definition 4.5: Fix n. Let $\Phi \in L^1_{loc}(\mathbb{R}^n)$ and assume that

$$(4.5) \qquad \Delta\tilde{\Phi} = \delta.$$

Then Φ is called a fundamental solution for the Laplacian in $\mathbb{R}^n$.

It is not clear a priori whether such a Φ exists. If it does exist it is far from unique. For let h be a function harmonic in $\mathbb{R}^n$. By (4.2), $\Delta\tilde{h} = 0$, whence

$$\Delta(\widetilde{\Phi+h}) = \delta.$$

Thus $\Phi + h$ again satisfies (4.5).

If we have a fundamental solution for a given n, we can use it to solve Poisson's equation

$$\Delta u = f$$

for any $f \in C_0^\infty(\mathbb{R}^n)$.

Theorem 4.2: Let Φ be a fundamental solution for the Laplacian in $\mathbb{R}^n$. Fix $f \in C_0^\infty(\mathbb{R}^n)$ and put

$$U(x) = \int \Phi(x-y)f(y)dy.$$

Then $\Delta U = f$.

Proof: A change of variable gives

$$U(x) = \int \Phi(y)f(x-y)dy.$$

Since $\Phi \in L_{loc}$ and $f \in C_0^\infty$, we get

$$\Delta U(x) = \int \Phi(y)\Delta_x(f(x-y))dy.$$

Now

$$\Delta_x(f(x-y)) = \Delta_y(f(x-y)),$$

so, fixing x,

$$\Delta U(x) = \tilde{\Phi}(\Delta_y g),$$

where $g(y) = f(x-y)$. But

$$\Delta U(x) = \tilde{\Phi}(\Delta g) = \Delta\tilde{\Phi}(g) = \delta(g) = g(0) = f(x).$$

Thus

$$\Delta U(x) = f(x). \qquad \text{Q.E.D.}$$

Because of (4.4), $-\frac{1}{4\pi}\frac{1}{|x|}$ is a fundamental solution for $\mathbb{R}^3$.

Fix n and look for a fundamental solution Φ for $\mathbb{R}^n$. Suppose Φ is smooth in some ball D, where $0 \notin D$. Choose $f \in C_0^\infty$ with $\operatorname{supp} f \subset D$. Then $f(0) = 0$, so $\delta(f) = 0$. So

$$0 = \delta(f) = \Delta\Phi(f) = \int \Delta f \Phi dx.$$

The last integral, since $\operatorname{supp} f \subset D$,

$$= \int f \Delta\Phi dx,$$

by G4. Since f was arbitrary, $\Delta\Phi = 0$ in D. Since D is arbitrary, we conclude that Φ is harmonic in $\mathbb{R}^n\setminus\{0\}$, provided Φ is smooth there.

We claim Φ must be singular at $x = 0$. For if not, it follows that Φ is harmonic in $\mathbb{R}^n$, and so by (4.2),

$$\Delta(\tilde{\Phi}) = \widetilde{\Delta\Phi} = 0,$$

contradicting $\Delta\tilde{\Phi} = \delta$.

Also, since δ is rotation-invariant, it is reasonable to look for a Φ that is rotation-invariant, i.e., for some function f on $\mathbb{R}$, $\Phi(x) = f(|x|)$. Put $r = |x|$. $\frac{\partial r}{\partial x_i} = \frac{x_i}{r}$. A brief calculation then gives:

$$\Delta\Phi = f'' + (n-1)\frac{1}{r}\cdot f' = \frac{1}{r^{n-1}}\,(f'(r)r^{n-1})'.$$

From the above, this implies that for $r > 0$, $(f'(r)r^{n-1})' = 0$, whence for $r > 0$:

(4.6) $$f'(r) = \frac{C}{r^{n-1}}, \quad C \text{ a constant.}$$

Let $n = 2$. Then $f(r) = C\log r$, so $\Phi(x) = C\log|x|$. We shall compute C.

Fix $\gamma \in C_0^\infty(\mathbb{R}^2)$. Choose R such that $\gamma = 0$ for $|x| > R - \epsilon$. Let Ω_ϵ be the region: $\epsilon < |x| < R$ in $\mathbb{R}^2$.

$$\int_{\partial\Omega_\epsilon}\left\{\log\frac{1}{|x|}\frac{\partial\gamma}{\partial n} - \gamma\frac{\partial}{\partial n}\left(\log\frac{1}{|x|}\right)\right\}ds = \iint_{\Omega_\epsilon}\log\frac{1}{|x|}\,\Delta\gamma\,dx.$$

Since $\frac{\partial\gamma}{\partial n} = \gamma = 0$ on $|x| = R$, the left side

$$= \int_{|x|=\epsilon}\log\frac{1}{\epsilon}\frac{\partial\gamma}{\partial n}\,ds - \int_{|x|=\epsilon}\gamma\cdot\frac{1}{\epsilon}\,ds.$$

Hence

$$\left(\log\frac{1}{\epsilon}\right)O(\epsilon) - \frac{1}{\epsilon}\,(2\pi\epsilon\cdot\gamma(0) + o(\epsilon)) = \iint_{\Omega_\epsilon}\log\frac{1}{|x|}\,\Delta\gamma\,dx.$$

Letting $\epsilon \to 0$, we get

(4.7) $$-2\pi\,\gamma(0) = \int\int_{|x|<R} \log\frac{1}{|x|}\,\Delta\gamma dx.$$

Put $\Phi(x) = -\frac{1}{2\pi}\log\frac{1}{|x|}$. (4.7) gives

$$\gamma(0) = \int_{\mathbb{R}^2} \Phi(x)\Delta\gamma\, dx = (\Delta\tilde{\Phi})(\gamma).$$

Thus we have

<u>Theorem 4.3</u>: $-\frac{1}{2\pi}\log\frac{1}{|x|}$ <u>is a fundamental solution in</u> $\mathbb{R}^2$.

<u>For</u> $n \neq 2$, (4.6) <u>gives</u> $f(r) = \frac{D}{r^{n-2}}$, <u>where</u> D <u>is a constant, and so</u>

$$\Phi(x) = D \cdot \frac{1}{|x|^{n-2}}.$$

<u>Exercise 4.1</u>: Let $n \neq 2$. Show that if σ_n denotes the (n-1)-dimensional area of the unit sphere in $\mathbb{R}^n$, then

$$\Phi(x) = -\frac{1}{\sigma_n}\cdot\frac{1}{n-2}\cdot\frac{1}{|x|^{n-2}}$$

is a fundamental solution in $\mathbb{R}^n$.

From this Exercise we at once get an n-dimensional analogue of Theorem 3.3: If $f \in C_0^\infty(\mathbb{R}^n)$, then for all $x \in \mathbb{R}^n$:

(4.8) $$f(x) = -\frac{1}{\sigma_n}\frac{1}{n-2}\int\frac{\Delta f(y)dy}{|x-y|^{n-2}}.$$

To see this, put $g(y) = f(x-y)$.

$$f(x) = g(0) = (\Delta\tilde{\Phi})(g) = \tilde{\Phi}(\Delta g) = C\int\frac{\Delta g(y)dy}{|y|^{n-2}},$$

where $C = -\frac{1}{\sigma_n} \cdot \frac{1}{n-2}$,

$$= C \int \frac{\Delta f(x-y)}{|y|^{n-2}}\, dy = C \int \frac{\Delta f(\zeta)}{|x-\zeta|^{n-2}}\, d\zeta,$$

which is (4.8).

Note: For all $n > 2$, the fundamental solution we have constructed

(i) is everywhere < 0 in $\mathbb{R}^n$

(ii) vanishes at ∞.

For $n = 2$, our fundamental solution

(i') is negative near $x = 0$, positive near $x = \infty$

(ii') becomes ∞ at ∞.

By Exercise 4.1, in $\mathbb{R}^n$, $n > 2$,

$$\Delta\left(\frac{1}{|x|^{n-2}}\right) = C\cdot\delta, \qquad C = -\sigma_n \cdot (n-2).$$

Thus $\frac{1}{|x|^{n-2}}$ plays the role in $\mathbb{R}^n$ that $\frac{1}{|x|}$ has in $\mathbb{R}^3$.

Definition 4.6: Fix $n > 2$. Let μ be a positive Borel measure on $\mathbb{R}^n$ of finite total mass. The potential of μ, denote U^μ, is given by

$$U^\mu(x) = \int \frac{d\mu(y)}{|x-y|^{n-2}} .$$

Note: The theory of these potentials in $\mathbb{R}^n$, $n > 3$, is virtually identical with the corresponding theory in $\mathbb{R}^3$. Because of the above-mentioned properties of the fundamental solution in $\mathbb{R}^2$, potential theory in $\mathbb{R}^2$, is somewhat different, and we shall henceforth stay in $\mathbb{R}^n$, $n \geq 3$.

Exercise 4.2: Calculate

$$\iiint\limits_{|\zeta|\leq 1} \frac{dV}{|x-\zeta|} \quad \text{for all } x \text{ in } \mathbb{R}^3.$$

Exercise 4.3: Show that the Laplacian in $\mathbb{R}^n$ commutes with rotations, i.e., show that if T is a rotation of $\mathbb{R}^n$ and f a function, then

$$\Delta(f \circ T) = (\Delta f) \circ T.$$

NOTE TO THE READER:

We shall normally state theorems of potential theory in $\mathbb{R}^n$, $n \geq 3$, and give the proofs in $\mathbb{R}^3$. The modifications needed for the general case are usually evident, and we leave them to the reader.

5. Capacity

Certain material objects have the property that electric charges move freely on them under the influence of an electric field. Such objects are called <u>conductors</u>.

Let B_1 be a conducting body surrounded by a conducting surface S_2, which is grounded. Initially S_2 is uncharged. We put a total charge e on B_1 and observe what happens.

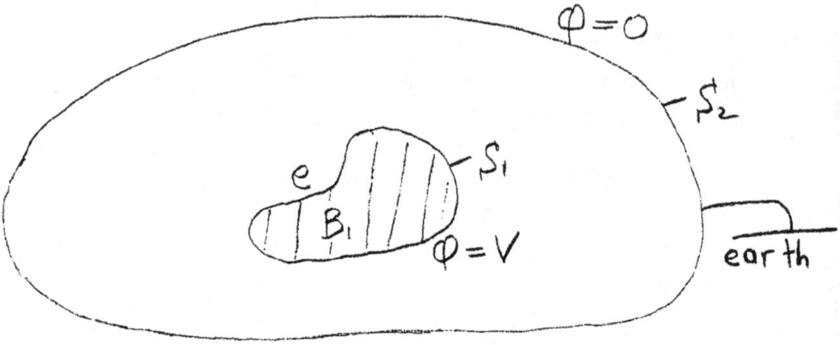

The charge on B_1 sets up an electric field which causes the charges on S_2 to move about. (One must regard S_2 which is "uncharged" to contain equal amounts of positive and negative charges.) After the charges on B_1 and S_2 have redistributed themselves so as to produce equilibrium (the net force on each charge is now 0), we obtain an electric field $\vec{\mathfrak{E}}$ in $\mathbb{R}^3$ with potential φ. It is observed that $\vec{\mathfrak{E}} = 0$ inside B_1 and that there are surface distributions of charge: $\omega_1 dS$ on S_1, the boundary of B_1, and $\omega_2 dS$ on S_2. We assume ω_1, ω_2 continuous on S_1, S_2.

$$\varphi(x) = \int_{S_1} \frac{\omega_1 dS}{|\zeta - x|} + \int_{S_2} \frac{\omega_2 dS}{|\zeta - x|} .$$

By Exercise 5.1 below, φ is continuous in $\mathbb{R}^3$.

(5.1) $\quad \varphi = 0$ on S_2, since S_2 is grounded.

Since $\vec{\mathfrak{E}} = -\text{grad}\,\varphi = 0$ inside B_1, φ is constant in B_1 and hence, by continuity, on S_1. Hence there is a constant V with

$$\varphi = V \quad \text{on} \quad S_1. \tag{5.2}$$

V depends on e. What happens to V if we change e?

Claim: V is proportional to e.

Proof: Let e' be some other charge on B_1, φ' the corresponding potential and V' the value of φ' on S_2.

Let Ω be the region between S_1 and S_2. Then $\frac{V'}{V}\cdot\varphi$ and φ' are two functions harmonic in Ω, continuous in $\bar{\Omega}$, $= 0$ on S_2 and $= V$ on S_1. It follows that

$$\varphi' = \frac{V'}{V}\varphi \quad \text{in} \quad \Omega. \tag{5.3}$$

Choose on S_1 the normal which points into Ω. Since φ is constant in B_1, $(\frac{\partial\varphi}{\partial n})^- = 0$ on S_1, so by Proposition 5.1[+]

$$\left(\frac{\partial\varphi}{\partial n}\right)^+ = \left(\frac{\partial\varphi}{\partial n}\right)^+ - \left(\frac{\partial\varphi}{\partial n}\right)^- = -4\pi\omega_1 \quad \text{on} \quad S_1.$$

The analogous relation holds for φ' and so

$$-4\pi\omega_1' = \left(\frac{\partial\varphi'}{\partial n}\right)^+ = \frac{V'}{V}\left(\frac{\partial\varphi}{\partial n}\right)^+ = -4\pi\frac{V'}{V}\omega_1.$$

Hence $\omega_1' = \frac{V'}{V}\omega_1$ and so

$$e' = \int_{S_1}\omega_1' dS = \frac{V'}{V}\int_{S_1}\omega_1 dS = \frac{V'}{V}e.$$

Thus

$$\frac{V'}{V} = \frac{e'}{e}, \quad \text{proving the Claim.}$$

[+] at the end of this section

Note: It follows that $\frac{e}{V}$ is a constant C depending only on B_1 and S_2. We have

(5.4) $$e = C \cdot V.$$

The number C is called the capacity of the condenser formed by S_1 and S_2. Since $\varphi = 0$ on S_2 and $\varphi = V$ on S_1, V is the potential difference between S_1 and S_2.

Formula (5.4) thus reads, in physical terms: The charge e stored in the condenser equals the product of the capacity of the condenser and the potential difference across the condenser.

Now fix S_1 and the charge e, but let S_2 move out to infinity. For each choice of S_2 we get a potential φ with

(i) φ harmonic between S_1 and S_2

(ii) φ = constant on S_1

(iii) $\varphi = 0$ on S_2.

If we are optimistic we may expect that, as $S_2 \to \infty$, φ will converge to a limiting function φ_∞. We should then expect the following properties from φ_∞:

(5.5) φ_∞ is harmonic in $\mathbb{R}^3$ outside S_1

(5.6) $\varphi_\infty = 0$ at ∞

(5.7) φ_∞ = constant on S_1.

We also expect that φ_∞ arises as the potential of distribution μ_∞ of the charge e on S_1, i.e.

(5.8) $$\varphi_\infty(x) = \int_{S_1} \frac{d\mu_\infty(\zeta)}{|x-\zeta|}, \qquad \int_{S_1} d\mu_\infty(\zeta) = e.$$

A function satisfying (5.5), (5.6), (5.7), (5.8) we shall call an equilibrium potential for B_1 corresponding to the charge e. We call μ_∞ the corresponding equilibrium distribution.

The problem of proving rigorously that such an equilibrium potential exists, under very general conditions, will occupy us in later sections. Let us show now that if $d\mu_\infty$ has a continuous surface density on S_1, the equilibrium potential is uniquely defined by (5.5), (5.6), (5.7), (5.8).

Let then φ, φ' be two functions satisfying these conditions, where (5.8) now means

$$(5.9) \qquad \varphi(x) = \int_{S_1} \frac{\omega dS}{|x-\zeta|}, \quad \int_{S_1} \omega dS = e$$

and

$$(5.9') \qquad \varphi'(x) = \int_{S_1} \frac{\omega' dS}{|x-\zeta|}, \quad \int_{S_1} \omega' dS = e.$$

Because of (5.7), there is a constant k with $\varphi' - k\varphi = 0$ on S_1. Since φ and φ' both vanish at ∞, it easily follows from the maximum principle for harmonic functions that $\varphi' - k\varphi = 0$ everywhere outside S_1. Since by (5.9) and (5.9') φ and φ' also are harmonic inside S_1, it similarly follows that $\varphi' - k\varphi = 0$ inside S_1. Hence on S_1,

$$\left(\frac{\partial\varphi'}{\partial n}\right)^+ - \left(\frac{\partial\varphi'}{\partial n}\right)^- = k\left(\left(\frac{\partial\varphi}{\partial n}\right)^+ - \left(\frac{\partial\varphi}{\partial n}\right)^-\right),$$

so by Proposition 5.1, $\omega' = k\omega$ on S_1. But

$$e = \int_{S_1} \omega' dS = k \int_{S_1} \omega dS = ke,$$

so $k = 1$. Thus $\varphi' = \varphi$, as claimed.

Let now B_1 be a body with boundary S_1. To each charge e on B_1 corresponds an equilibrium potential φ_∞ taking a constant value V_∞ on S_1. We see, as earlier for condensers, that V_∞ is proportional to e, i.e. $\exists$ constant C_∞ such that

$$(5.10) \qquad e = C_\infty \cdot V_\infty .$$

Thus C_∞ equals the amount of charge on B_1 which, when distributed so as to produce equilibrium, produces a potential difference of 1 between S_1 and ∞.

We call C_∞ the <u>capacity of</u> B_1.

Example: Let B_1 be the ball $\{x\ |x| \leq a\}$. Fix the charge e on B_1. For reasons of symmetry we expect the equilibrium distribution μ_∞ to be uniform with respect to area on ∂B_1, i.e., $\mu_\infty = \omega dS$, where ω is a constant such that

$$\int_{\partial B_1} \omega dS = 4\pi a^2 \omega = e \quad \text{or} \quad \omega = \frac{e}{4\pi a^2} .$$

The equilibrium potential φ is then

$$\varphi(x) = \omega \int_{|\zeta|=a} \frac{dS}{|x-\zeta|} .$$

We expect φ to be rotation invariant. In fact, let T be a rotation

$$\varphi(Tx) = \omega \int_{|\zeta|=a} \frac{dS}{|Tx-\zeta|} = \omega \int_{|\zeta|=a} \frac{dS}{|x-T^{-1}\zeta|} = \omega \int \frac{dS}{|x-\zeta|} ,$$

since T leaves dS invariant. So $\varphi(Tx) = \varphi(x)$. Hence $\exists$ function f of one variable with $\varphi(x) = f(|x|)$.

By a calculation we made in the last Section, $\Delta\varphi = \frac{1}{r^2} (f'(r)r^2)'$. Since φ is harmonic inside and outside $|x| = a$, we get

$$f'(r)r^2 = \begin{cases} A, & r > a \\ B, & r < a, \end{cases}$$

where A,B are constants. Hence

$$f(r) = \begin{cases} -\frac{A}{r} + A_1, & r > a \\ -\frac{B}{r} + A_2, & r < a . \end{cases}$$

$\varphi(\infty) = 0$, so $A_1 = 0$. φ is continuous at 0, hence $B = 0$, and on $|x| = a$, so $-\frac{A}{a} = A_2$. Thus

$$f(r) = \begin{cases} -\frac{A}{r}, & r > a \\ -\frac{A}{a}, & r < a \end{cases},$$

or

$$\varphi(x) = \begin{cases} \frac{c}{|x|}, & |x| > a \\ \frac{c}{a}, & |x| < a \end{cases}$$

where c is a constant. Also

$$\varphi(0) = \omega \int_{|\zeta|=a} \frac{dS}{|\zeta|} = \frac{e}{4\pi a^2} \cdot \frac{4\pi a^2}{a} = \frac{e}{a}$$

$\therefore$ $c = e$ and we have at last

(5.11) $$\varphi(x) = \frac{e}{|x|}, \quad |x| \geq a, \quad \varphi(x) = \frac{e}{a}, \quad |x| \leq a.$$

From (5.11), we verify that φ has the defining properties of an equilibrium potential for B_1.

Since $\varphi = \frac{e}{a}$ on ∂B_1, (5.10) yields that the capacity of $B_1 = a$.

Thus: the capacity of a ball equals its radius.

This result has physical meaning related to condensers, as follows: Let a condenser consist of a conducting ball: $|x| \leq a$ inside a conducting shell: $|x| = b$, $b > a$. To calculate its capacity, we need a function φ_b harmonic in $a < |x| < b$ with $\varphi_b = 0$ on $|x| = b$, $\varphi_b = V$ on $|x| \leq a$, where V is a constant, such that on $|x| = a$,

$$\left(\frac{\partial\varphi_b}{\partial n}\right)^+ = \left(\frac{\partial\varphi_b}{\partial n}\right)^+ - \left(\frac{\partial\varphi_b}{\partial n}\right)^- = -4\pi\omega,$$

where ω, as before, $= \frac{e}{4\pi a^2}$. We put

$$\varphi_b(x) = \frac{\alpha}{|x|} + \beta, \quad |x| > a$$

and seek constants α, β to satisfy our conditions

$$\left(\frac{\partial\varphi_b}{\partial n}\right)^+ = -\frac{\alpha}{|x|^2} = -\frac{\alpha}{a^2} \quad \text{on} \quad |x| = a,$$

so

$$-\frac{\alpha}{a^2} = -4\pi \frac{e}{4\pi a^2}, \quad \text{so} \quad \alpha = e.$$

To have $\varphi_b = 0$ on $|x| = b$, we then need $\beta = -\frac{e}{b}$. Thus

$$\varphi_b(x) = \frac{e}{|x|} - \frac{e}{b}.$$

So $V = \frac{e}{a} - \frac{e}{b}$ and the capacity C of the condenser is given by

$$C = \frac{e}{V} = \frac{e}{\frac{e}{a} - \frac{e}{b}} = \frac{ab}{b-a}.$$

For large b, C is thus near a. We interpret this as saying that the capacity of a ball is the limit of the capacities of condensers consisting of the ball and a conducting shell S_2, as $S_2 \to \infty$.

In Section 7 we shall give a mathematical definition of the "capacity" of an arbitrary Borel subset of $\mathbb{R}^n$.

We need some results on potentials of surface distribution in $\mathbb{R}^3$.

Let Σ be a bounded region on a smooth surface in $\mathbb{R}^3$, x_0 a point of Σ and ξ^+ and ξ^- the two rays normal to Σ at x_0. Let $\frac{\partial}{\partial n}$ denote the directional derivative in the direction of ξ^+.

For F a function defined in a neighborhood of x_0 in $\mathbb{R}^3$, we write

$(\frac{\partial F}{\partial n})^+(x_0)$

for the limit of $\frac{\partial F}{\partial n}(x)$ as $x \to x_0$ along ξ^+. We define $(\frac{\partial F}{\partial n})^-(x_0)$ similarly. Of course, if F is smooth in a neighborhood of x_0, then

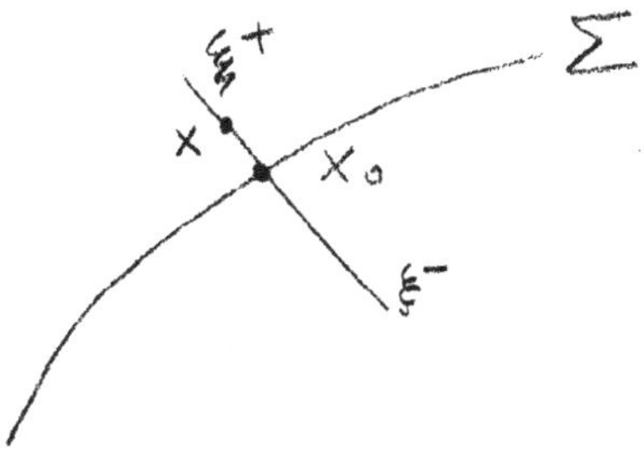

$$(\frac{\partial F}{\partial n})^+(x_0) = (\frac{\partial F}{\partial n})^-(x_0) = \frac{\partial F}{\partial n}(x_0).$$

Let now ρ be a smooth function defined on Σ. Put

$$U(x) = \int_\Sigma \frac{\rho(\zeta)dS}{|x-\zeta|} .$$

Exercise 5.1: U is continuous at each point of Σ.

By contrast, the partial derivatives of U are discontinuous across Σ.

Proposition 5.1: Let Σ, x_0, ρ, U be as above. Fix a normal ray ξ^+ to Σ at x_0. Let $\frac{\partial}{\partial n}$ be the corresponding directional derivative. Then

$$(\frac{\partial U}{\partial n})^+(x_0) - (\frac{\partial U}{\partial n})^-(x_0) = 4\pi\rho(x_0).$$

The proof of this result is elementary but involved. We refer the reader to Kellogg's book, [9] in Section 1 above, p. 164.

6. Energy

In the last section we were led to the notion of equilibrium potential φ_∞ for a body B in $\mathbb{R}^3$, where φ_∞ satisfies (5.5), (5.6), (5.7), (5.8). This potential arose as follows: we put a system of positive charges with total charge e, on a conductor B and let them distribute themselves on B until they reach equilibrium. The resulting distribution μ_∞ we called the equilibrium distribution on B corresponding to e and

$$\varphi_\infty = U^{\mu_\infty}.$$

Every distribution of charges on B has a certain potential energy and we expect equilibrium to occur for that distribution of the given charge which minimizes the potential energy.

To utilize this idea in proving the existence of an equilibrium distribution, we need a mathematical expression for the potential energy of a charge distribution μ.

We derive such an expression first for a finite system of point charges.

Let $x_1,\ldots,x_n$ be points in $\mathbb{R}^3$ and consider charges e_i located at x_i, $i = 1,\ldots,n$. Since the e_i repel each other, to keep e_i at x_i we must imagine it nailed down there. If we pull out all the nails, the charges will fly apart, thereby doing work. The total work they are able to do in this way is called the potential energy of the system of charges. We compute this work as follows: Keeping $e_1,\ldots,e_{n-1}$ fixed, release e_n. e_n moves to infinity due to the force exerted by the other e_i, this force being $-e_n \operatorname{grad} U$, where

$$U(x) = \sum_{i=1}^{n-1} \frac{e_i}{|x-x_i|} .$$

Let $x = x(s)$ be the path taken by e_n, where $x(0) = x_n$, $x(\infty) = \infty$ and s is arc length. The work W done in this motion is given by

$$W_n = \int_0^\infty (-e_n \operatorname{grad} U) \cdot \frac{dx}{ds}\, ds$$

$$= -\int_0^\infty e_n \frac{dU}{ds}\, ds = e_n U(x_n)$$

$$= \sum_{i=1}^{n-1} \frac{e_i e_n}{|x_n - x_i|} .$$

We next release e_{n-1} and calculate the corresponding work similarly as

$$W_{n-1} = \sum_{i=1}^{n-2} \frac{e_i e_{n-1}}{|x_{n-1} - x_i|} .$$

The total work W done when all charges are released, and hence the potential energy of the original system, is given by

$$W = \sum_1^n W_i = \sum_{i<j} \frac{e_i e_j}{|x_i - x_j|} ,$$

or

(6.1) $$W = \frac{1}{2} \sum_{\substack{i,j \\ i \neq j}} \frac{e_i e_j}{|x_i - x_j|} .$$

By analogy to (6.1) we define for each measure μ on $\mathbb{R}^3$ the <u>energy of</u> μ, denoted $I(\mu)$, by

(6.2) $$I(\mu) = \int \int \frac{d\mu(x) d\mu(y)}{|x-y|} .$$

(We suppress the factor $\frac{1}{2}$ from (6.1).)

Of course $I(\mu) = +\infty$ for some measures. As we said earlier, we expect that the distribution $\overline{\mu}$ on B with $\int_B d\overline{\mu} = e$ which minimizes $I(\mu)$, i.e., such that

(6.3) $I(\overline{\mu}) \leq I(\mu)$ for all measures μ on B with $\int d\mu = e$, will be the equilibrium distribution.

We have two tasks: First, to show ∃ a measure $\bar{\mu}$ on B satisfying (6.3), and then to show that $U^{\bar{\mu}}$ satisfies conditions (5.5), (5.6), (5.7) and (5.8).

For $n > 3$ and μ a measure on $\mathbb{R}^n$ we define

$$(6.4) \qquad I(\mu) = \int \int \frac{d\mu(x)d\mu(y)}{|x-y|^{n-2}} .$$

<u>Proposition 6.1: Let</u> μ <u>be a measure on</u> $\mathbb{R}^n$, $n \geq 3$,

$$(6.5) \qquad I(\mu) = \int U^{\mu}(x)d\mu(x).$$

<u>Note:</u> The proof of (6.5) is immediate. (6.5) shows that, in order for μ to have finite energy it suffices that

(a) μ has compact support, and

(b) U^{μ} is bounded on $\mathbb{R}^n$.

For μ a finite sum of point masses, $I(\mu)$ is clearly infinite, on the other hand.

For a special case, we can get an interesting expression for the energy.

<u>Theorem 6.2: Let</u> $\mu = pdx$, <u>where</u> $p \in C_0^2(\mathbb{R}^3)$ <u>and put</u> $U = U^{\mu}$. <u>Then</u>

$$(6.9) \qquad I(\mu) = \frac{1}{4\pi} \int_{\mathbb{R}^3} \left\{ \left(\frac{\partial U}{\partial x_1}\right)^2 + \left(\frac{\partial U}{\partial x_2}\right)^2 + \left(\frac{\partial U}{\partial x_3}\right)^2 \right\} dx_1 dx_2 dx_3 .$$

<u>Note:</u> In (6.9) we even allow p to change sign, so that μ is merely a signed measure.

<u>Proof:</u> We claim for large $|x| = r$

(6.10) $$U = O(\tfrac{1}{r}) \quad \text{and} \quad \frac{\partial U}{\partial r} = O(\tfrac{1}{r^2}).$$

For fix ξ.

$$\frac{\partial}{\partial x_i}\left(\frac{1}{|x-\xi|}\right) = \frac{-(x_i-\xi_i)}{|x-\xi|^3}.$$

$\frac{\partial}{\partial r} = \sum_{i=1}^{3} A_i \frac{\partial}{\partial x_i}$ where $|A_i| \leq 1$, all i. So

$$\frac{\partial}{\partial r}\left(\frac{1}{|x-\xi|}\right) \leq \frac{3}{|x-\xi|^2}.$$

$$U(x) = \int \frac{d\mu(\xi)}{|x-\xi|}, \quad \frac{\partial U}{\partial r}(x) = \int \frac{\partial}{\partial r}\left(\frac{1}{|x-\xi|}\right) d\mu(\xi), \quad \left|\frac{\partial U}{\partial r}(x)\right| \leq \int \frac{3}{|x-\xi|^2}\, d\mu(\xi).$$

Let $d = \max |\xi|$ for $\xi \in \operatorname{supp} \mu$. $|x-\xi| \geq |x| - |\xi| \geq |x| - d$, for $\xi \in \operatorname{supp} \mu$. So

$$\left|\frac{\partial U}{\partial r}(x)\right| \leq \frac{3}{(r-d)^2}\|\mu\| \leq \frac{4\|\mu\|}{r^2},$$

for large r. Similarly $U = O(\frac{1}{r})$ and (6.10) holds. Fix R large. By G.3] applied to the region $|x| < R$, we have

(6.11) $$\int_{|x|=R} U \frac{\partial U}{\partial r}\, dS = \int_{|x|<R} \operatorname{grad} U \cdot \operatorname{grad} U dx + \int_{|x|<R} U \Delta U dx.$$

Recall that $\Delta U = -4\pi\rho$, and let $R \to \infty$ in (6.11). The right side $\to$

$$\int_{\mathbb{R}^3} |\operatorname{grad} U|^2 dx - 4\pi \int_{\mathbb{R}^3} U \cdot \rho dx.$$

$$\left|\int_{|x|=R} U \frac{\partial U}{\partial r}\, dS\right| = O(\tfrac{1}{R}) O(\tfrac{1}{R^2}) 4\pi R^2,$$

and so $\to 0$ as $R \to \infty$. Hence, recalling (6.5) we conclude

$$0 = \int_{\mathbb{R}^3} |\text{grad } U|^2 dx - 4\pi I(\mu), \text{ q.e.d.}$$

Theorem 6.3: Let E be a compact set in $\mathbb{R}^n$. Among all probability measures on E there is one which minimizes the energy, i.e. $\exists\, \bar{\mu}$ on E, $\bar{\mu}(E) = 1$ such that

$$I(\bar{\mu}) \leq I(\mu) \tag{6.12}$$

for all μ on E with $\mu(E) = 1$.

We need:

Lemma 6.4: Given a compact set E in $\mathbb{R}^n$ and a sequence of probability measures $\{\mu_n\}$ on E converging weakly to $\bar{\mu}$. Then the sequence $\{\mu_n \times \mu_n\}$ of product measures on $E \times E$ converges weakly to $\bar{\mu} \times \bar{\mu}$.

Proof: Let $f, g \in C(E)$. Then $f(x)\cdot g(y) \in C(E \times E)$. Hence

$$\int f(x)g(y)d(\mu_n \times \mu_n) = \int f d\mu_n \cdot \int g d\mu_n \to \int f d\bar{\mu} \cdot \int g d\bar{\mu} = \int f(x)g(y)d(\bar{\mu} \times \bar{\mu}).$$

It follows that

$$\int F d(\mu_n \times \mu_n) \to \int F d(\bar{\mu} \times \bar{\mu}) \tag{6.13}$$

whenever F is a finite sum of functions $f_i(x)g_i(y)$, $f_i, g_i \in C(E)$. By the Weierstrass approximation theorem on $E \times E$ such sums are dense in $C(E \times E)$. It follows that (6.13) holds for all $F \in C(E \times E)$. q.e.d.

Definition 6.1: If F is a compact set in $\mathbb{R}^n$, $P(F)$ denotes the set of all probability measures on F.

Proof of Theorem 6.3: Put

$$\gamma = \inf_{\mu \in P(E)} I(\mu).$$

Choose $\{\mu_n\} \in P(E)$ with $I(\mu_n) \to \gamma$. $\exists$ subsequence, again denoted $\{\mu_n\}$, converging weakly to $\bar{\mu} \in P(E)$.

We claim $I(\bar{\mu}) = \gamma$.

$$I(\mu_n) = \int_{E \times E} \frac{d\mu_n(x)d\mu_n(y)}{|x-y|} .$$

Since $\mu_n \to \bar{\mu}$ weakly, $\mu_n \times \mu_n \to \bar{\mu} \times \bar{\mu}$ weakly on $E \times E$, by the preceding Lemma, i.e.,

$$\int_{E \times E} K(x,y)d\mu_n(x)d\mu_n(y) \to \int_{E \times E} K(x,y)d\bar{\mu}(x)d\bar{\mu}(y)$$

whenever $K \in C(E \times E)$. Unfortunately $\frac{1}{|x-y|} \notin C(E \times E)$.

Put

$$K_j(t) = \begin{cases} \frac{1}{t}, & t > \frac{1}{j} \\ j, & 0 \le t \le \frac{1}{j} \end{cases} .$$

K_j is continuous, so $K_j(|x-y|) \in C(E \times E)$. Hence, if for each μ we put

$$I_j(\mu) = \int_{E \times E} K_j(|x-y|)d\mu(x)d\mu(y),$$

$I_j(\mu_n) \to I_j(\bar{\mu})$ as $n \to \infty$, j fixed.

Also, $I_j(\mu) \le I(\mu)$ for all μ, so

$$\lim_{n\to\infty} I_j(\mu_n) \leq \lim_{n\to\infty} I(\mu_n),$$

$$I_j(\overline{\mu}) \leq \gamma, \text{ i.e.,}$$

(6.14) $$\int_{E\times E} K_j(|x-y|)d(\overline{\mu}\times\overline{\mu}) \leq \gamma.$$

Also, as $j\to\infty$, $K_j(|x-y|)$ increases at each (x,y) to $\frac{1}{|x-y|}$. Hence by the monotone convergence theorem, (6.14) yields

$$\int_{E\times E} \frac{1}{|x-y|}\, d(\overline{\mu}\times\overline{\mu}) \leq \gamma,$$

or $I(\overline{\mu}) \leq \gamma$. Hence $I(\overline{\mu}) = \gamma$. q.e.d.

7. Existence of the Equilibrium Distribution

In Section 5 we introduced the notion of capacity of a body B in $\mathbb{R}^3$ as follows: the capacity of B equals the amount of charge which, when distributed on B so as to produce equilibrium, yields a potential which is identically 1 on B. In other words, choose a measure μ_0 on B such that $U^{\mu_0} = 1$ on B. Then the capacity of $B = \mu_0(B)$.

Now observe: if μ is any measure on B with $U^\mu \leq 1$ on B, then

$$\mu(B) = \int U^{\mu_0} d\mu = \int \left\{ \int \frac{d\mu_0(y)}{|x-y|} \right\} d\mu(x) = \int \left(\frac{d\mu(x)}{|x-y|} \right) d\mu_0(y)$$

$$= \int U^\mu d\mu_0 \leq \mu_0(B).$$

Hence $\mu_0(B) = \sup \mu(B)$ taken over all measures μ on B with $U^\mu \leq 1$ on B.

This relation leads us to a new definition of capacity, which does not require us to know that an equilibrium distribution exists, to wit:

<u>Definition 7.1</u>: Let E be a bounded Borel set in $\mathbb{R}^n$. The <u>capacity of</u> E, $C(E) = \sup \mu(E)$ taken over all measures μ with $\operatorname{supp} \mu \subseteq E$ such that $U^\mu \leq 1$ on E.

<u>Note</u>: If $\mu = 0$ is the only measure with support in E such that $U^\mu \leq 1$, then $C(E) = 0$. This happens, for instance, if E is a finite set.

On the other hand, if E admits a measure ν with bounded potential U^ν and $\nu \neq 0$, then $C(E) > 0$. For if $M > U^\mu$, then putting $\nu' = \frac{\nu}{M}$ we have $U^{\nu'} \leq 1$, so $C(E) \geq \nu'(E) > 0$. Thus: a set E has positive capacity if and only $\exists$ measure $\mu \neq 0$ with $\operatorname{supp} \mu \subseteq E$ and U^μ bounded.

The sets of capacity 0 are the "small" sets of potential theory, playing a role somewhat analogous to sets of Lebesgue measure zero in integration theory. However, as we shall see, capacity is not itself a measure, i.e., is not additive on disjoint sets. The main result of this section is this:

Theorem 7.1: Let E be a compact set $\subset \mathbb{R}^n$. Let $\bar{\mu}$ be a probability measure on E which minimizes energy, as given by Theorem 6.3, and let $I(\bar{\mu}) = \gamma$. Then $U^{\bar{\mu}} = \gamma$ on E, except possibly on a subset of E of capacity 0. Also $U^{\bar{\mu}} \leq \gamma$ everywhere in $\mathbb{R}^n$.

Note: $\gamma = \infty$ only if $C(E) = 0$, in which case the Theorem has no content. So without loss of generality, we assume $\gamma < \infty$.

We proceed to interpret the energy as the squared norm in a certain inner product on a vector space.

Definition 7.2: Let α,β be measures on $\mathbb{R}^n$.

$$[\alpha,\beta] = \int \frac{d\alpha(x)d\beta(y)}{|x-y|^{n-2}} . \qquad (7.1)$$

Note: $[\alpha,\beta]$ is well-defined and ≥ 0, but may $= +\infty$.

Fix a compact set E in $\mathbb{R}^n$. Consider the vector space of all signed measures $\nu = \alpha - \beta$, α,β measures on E. Each such ν is a real-valued set function defined on the Borel sets $\subset E$. On this vector space, $[\ ,\ ]$ as in (7.1), is formally an inner product, except for the condition $[\nu,\nu] > 0$ if $\nu \neq 0$, which is in no way clear.

Also, for μ a measure on E, $[\mu,\mu] = I(\mu)$, so that $I(\mu)$ is the squared norm for this inner product.

$P(E)$, the set of probability measures on E, is a convex set in our vector space and the measure $\bar{\mu}$ is an element of $P(E)$ of minimal norm, $\|\bar{\mu}\|^2 = I(\bar{\mu}) = \gamma$.

Consider the functional L, $L(\tau) = [\bar{\mu},\tau] - [\bar{\mu},\bar{\mu}]$. We have the following picture:

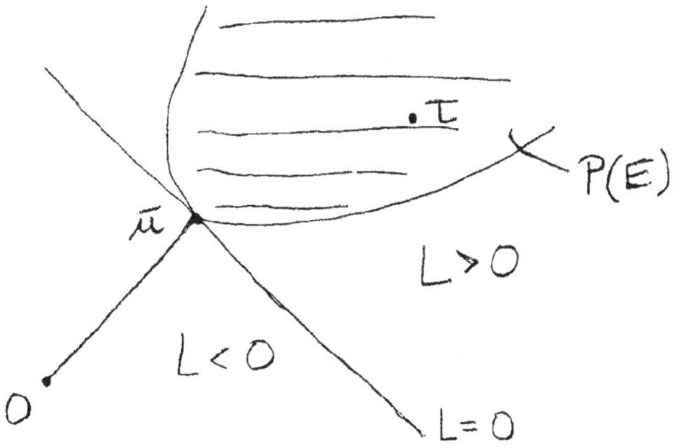

This picture suggests that P(E) lies entirely in the set: $L \geq 0$, i.e., that $\tau \in P(E)$ implies $[\bar{\mu},\tau] \geq [\bar{\mu},\bar{\mu}] = \gamma$.

Assertion 1: If $\tau \in P(E)$, $I(\tau) < \infty$, then $[\bar{\mu},\tau] \geq [\bar{\mu},\bar{\mu}] = \gamma$.

Proof: Fix $\tau \in P(E)$, $I(\tau) < \infty$. Put $\mu_\delta = (1-\delta)\bar{\mu} + \delta\tau$, where $0 < \delta < 1$. Then $\mu_\delta \in P(E)$ and so $I(\mu_\delta) \geq \gamma$. Thus

$$\begin{aligned}\gamma &\leq [(1-\delta)\bar{\mu} + \delta\tau,\ (1-\delta)\bar{\mu} + \delta\tau] \\ &= (1-\delta)^2\gamma + \delta^2[\tau,\tau] + 2\delta(1-\delta)[\bar{\mu},\tau] \\ &= \gamma - 2\delta\gamma + 2\delta[\bar{\mu},\tau] + o(\delta^2).\end{aligned}$$

Hence

$$0 \leq 2\delta\{[\bar{\mu},\tau] - \gamma\} + o(\delta^2),$$

whence $[\bar{\mu},\tau] \geq \gamma$, as asserted.

We shall deduce that $U^{\bar{\mu}} \geq \gamma$ on E except for a set of capacity 0. Put

$$T = \{x \in E | U^{\bar{\mu}}(x) < \gamma\}.$$

Exercise 7.1: T is a Borel set.

Assertion 2: $C(T) = 0$.

Proof: Suppose $C(T) > 0$. Then $\tau \in P(E)$, $\operatorname{supp} \tau \subset E$, U^{τ} bounded and so $I(\tau) < \infty$. By Assertion 1,

$$\int \overline{U^{\mu}} d\tau \geq \gamma .$$

Also $\overline{U^{\mu}} < \gamma$ on T, so

$$\int \overline{U^{\mu}} d\tau < \gamma \int d\tau = \gamma ,$$

which is a contradiction. Hence $C(T) = 0$, as asserted.

Lemma 7.2: Let μ be a measure on $\mathbb{R}^n$ of compact support E. Then

$$\sup_{\mathbb{R}^n} U^{\mu} \leq 2 \cdot \sup_{E} U^{\mu} . \tag{7.2}$$

Proof: Fix $x \in \mathbb{R}^3$ and let ζ be the nearest point to x on E. If $y \in E$,

$$|\zeta - y| \leq |\zeta - x| + |x - y| ,$$

so

$$\frac{|\zeta - y|}{|x - y|} \leq \frac{|\zeta - x|}{|x - y|} + 1 \leq 2 .$$

Hence

$$U^{\mu}(x) = \int \frac{d\mu(y)}{|x-y|} \leq 2 \int \frac{d\mu(y)}{|\zeta - y|} = 2U^{\mu}(\zeta) ,$$

whence (7.2).

Lemma 7.3: Let S be a Borel set in $\mathbb{R}^k$ with $C(S) = 0$. Then for each measure α on $\mathbb{R}^k$ with $I(\alpha) < \infty$, $\alpha(S) = 0$.

Proof: Put $E_n = \{x \in S \mid U^\alpha(x) < n\}$ and $E_\infty = \{x \in S \mid U^\alpha(x) = \infty\}$. Since by hypothesis $I(\alpha) = \int U^\alpha d\alpha < \infty$, $\alpha(E_\infty) = 0$. If $\alpha(E_n) = 0$ for every n, then $\alpha(S) = 0$.

If not, then $\alpha(E_n) > 0$ for some n. Choose F_n closed $\subseteq E_n$ with $\alpha(F_n) > 0$ and set $\alpha' = \alpha|_{F_n}$. Then $\operatorname{supp} \alpha' \subseteq E_n$ and $\alpha' \leq \alpha$. Hence $U^{\alpha'} \leq U^\alpha \leq n$ on $\operatorname{supp} \alpha'$. Lemma 7.2 gives that $U^{\alpha'} \leq 2n$ everywhere. Since $\operatorname{supp} \alpha' \subseteq S$, we conclude $C(S) > 0$, which is false.

Hence $\alpha(S) = 0$. q.e.d.

We now state an improved version of (7.2), and prove it in the next section.

Theorem 7.4: Let μ be a measure of compact support E. Then

$$\sup_{\mathbb{R}^n} U^\mu \leq \sup_E U^\mu . \tag{7.3}$$

Note: If U^μ happens to be continuous on $\mathbb{R}^n$, (7.3) is immediate, since U^μ is harmonic on $\mathbb{R}^n \setminus E$ and vanishes at ∞, so for each $x \in \mathbb{R}^n$

$$U^\mu(x) \leq \max_{\partial E} U^\mu .$$

However, in general U^μ is not continuous and (7.3) is not at all obvious.

Lemma 7.5: Let μ be a measure on $\mathbb{R}^n$ of compact support and fix $x_0 \in \mathbb{R}^n$. If $U^\mu(x_0) > a$, then $U^\mu(x) > a$ for all x in some neighborhood of x_0. In other words: U^μ is lower semi-continuous at each point.

Proof: As earlier, put

$$K_j(t) = \begin{cases} \frac{1}{t}, & t > \frac{1}{j} \\ j, & t \leq \frac{1}{j} \end{cases}$$

$$U_j(x) = \int K_j(|x-y|)d\mu(\zeta) \to U^{\mu}(x) \quad \text{as} \quad j \to \infty$$

for each x. Hence $U_j(x_0) > a$ for $j > j_0$. U_j is continuous, so $U_j(x) > a$ for $|x-x_0| < \delta$ for some δ. But $U^{\mu} \geq U_j$, so $U^{\mu}(x) > 0$ for $|x-x_0| < \delta$. q.e.d.

In the notation of Theorem 7.1, we have

Assertion 3: $U^{\bar{\mu}} = \gamma$ a.e. $-d\bar{\mu}$.

Proof: $T = \{x \in E | U^{\bar{\mu}}(x) < \gamma\}$. By Assertion 2, $C(T) = 0$. By Lemma 7.3, this gives $\bar{\mu}(T) = 0$. So

$$\gamma = \int_E U^{\bar{\mu}} d\bar{\mu} = \int_{E \setminus T} U^{\bar{\mu}} d\bar{\mu}.$$

Also $\int_{E \setminus T} d\bar{\mu} = 1$. So

$$0 = \int_{E \setminus T} (U^{\bar{\mu}} - \gamma) d\bar{\mu}.$$

But $U^{\bar{\mu}} - \gamma \geq 0$ on $E \setminus T$. Hence $U^{\bar{\mu}} - \gamma = 0$ a.e. $-d\bar{\mu}$, as claimed.

Proof of Theorem 7.1: Suppose $\exists x_0 \in \operatorname{supp} \bar{\mu}$ with $U^{\bar{\mu}}(x_0) > \gamma$. By Lemma 7.5, there then exists a neighborhood $\mathfrak{N}$ of x_0 where $U^{\bar{\mu}} > \gamma$. Since $x_0 \in \operatorname{supp} \bar{\mu}$, $\bar{\mu}(\mathfrak{N}) > 0$, and this contradicts Assertion 3. Hence $U^{\bar{\mu}} \leq \gamma$ on $\operatorname{supp} \bar{\mu}$. By (7.3) this give $U^{\bar{\mu}} \leq \gamma$ everywhere. By Assertion 2, the subset of E where $U^{\bar{\mu}} < \gamma$ has capacity 0. Hence $U^{\bar{\mu}} = \gamma$ outside a set of capacity 0. Theorem 7.1 is proved.

How "small" is a set of capacity 0?

Exercise 7.2: If S is a Borel set in $\mathbb{R}^n$ and $C(S) = 0$, then $dx(S) = 0$, i.e., the n-dimensional Lebesgue measure of S is 0.

Problem: Theorem 7.1 leaves us with an exceptional set, the subset $\tilde{E}$ of E where $U^{\bar{\mu}} < \gamma$. We know $C(\tilde{E}) = 0$. What else can be said about $\tilde{E}$?

Electrostatics suggests that $U^{\bar{\mu}} = \gamma$ on int E, so that we expect that $\tilde{E} \subset \partial E$. We also expect that $\bar{\mu}$ lies on ∂E. In fact, one can prove:

Theorem 7.1a: $\tilde{E} \subset \partial E$; $\bar{\mu}$ lies on ∂E; $U^{\bar{\mu}} = \gamma$ on int E.

Proof: Fix an open ball $B \subset$ int E. $C(\tilde{E}) = 0$, so by Exercise 7.2, $dx(\tilde{E}) = 0$, so $dx(B \cap \tilde{E}) = 0$. Thus $U^{\bar{\mu}} = \gamma$ a.e. $-dx$ in B.

Choose $f \in C_0^2(B)$. Then

$$\gamma \int_B \Delta f dx = \int_B U^{\bar{\mu}}(x) \Delta f dx$$

$$= \int_B \left\{ \int \frac{d\bar{\mu}(\zeta)}{|x-\zeta|} \right\} \Delta f(x) dx$$

$$= \int d\bar{\mu}(\zeta) \int_B \frac{\Delta f(x)}{|x-\zeta|} dx = -4\pi \int_B f(\zeta) d\bar{\mu}(\zeta).$$

Also, by G.4

$$\int_B (1\Delta f - f\Delta 1) dx = \int_{\partial B} \left(1 \frac{\partial f}{\partial n} - f \frac{\partial 1}{\partial n}\right) dS, \quad \text{or}$$

$$\int_B \Delta f dx = 0.$$

Thus

Thus

$$\int_B f(\zeta)d\bar{\mu}(\zeta) = 0.$$

Since f was arbitrary in $C_0^2(B)$, $\bar{\mu}(B) = 0$. It follows that $\bar{\mu}$ vanishes on int E, so $\bar{\mu}$ lies on ∂E, as asserted. This gives that $U^{\bar{\mu}}$ is continuous on int E. But $U^{\bar{\mu}} = \gamma$ on int E, outside of a set of Lebesgue measure 0, hence on a dense set. So $U^{\bar{\mu}} = \gamma$ everywhere on int E. We are done.

<u>Note</u>: A natural question, which we will tackle in Section 10, is to find conditions on E under which $\tilde{E}$ is empty.

<u>Theorem 7.1b</u>: <u>Let</u> $x_0 \in E\backslash\tilde{E}$. <u>Then</u> $U^{\bar{\mu}}$ <u>is continuous at</u> x_0.

<u>Proof</u>: Since $x_0 \in E\backslash\tilde{E}$, $U^{\bar{\mu}}(x_0) = \gamma$. Fix $\epsilon > 0$. By Lemma 7.5 $\exists$ neighborhood $\mathfrak{N}$ of x_0 with $U^{\bar{\mu}} > \gamma - \epsilon$ in $\mathfrak{N}$.

Also $U^{\bar{\mu}} \leq \gamma$ in $\mathfrak{N}$, since this inequality holds everywhere. Thus

$$|U^{\bar{\mu}} - U^{\bar{\mu}}(x_0)| < \epsilon \quad \text{for } x \text{ in } \mathfrak{N}. \quad \text{q.e.d.}$$

<u>Note</u>: $\bar{\mu}$ is the equilibrium distribution of the total mass 1 on E, since $\bar{\mu}(E) = 1$.

Given an arbitrary number $e > 0$, we have a corresponding equilibrium distribution of total mass e, namely $e\bar{\mu}$. The corresponding potential

$$U^{e\bar{\mu}} = e\gamma \quad \text{on } E\backslash\tilde{E}.$$

In particular, putting $e = \frac{1}{\gamma}$, we have

<u>Theorem 7.1c</u>: <u>Put</u> $\mu_E = \frac{1}{\gamma}\bar{\mu}$. <u>Then</u> μ_E <u>is a measure on</u> E <u>such that</u>

(7.4) $$U^{\mu_E} = 1 \text{ \underline{on} } E, \text{ \underline{except on} } \tilde{E}.$$

(7.5) $$\frac{1}{\gamma} = \mu_E(E) = C(E).$$

The proof of (7.5) is left as an exercise.

We conclude this section with some general properties of capacity.

Exercise 7.3: If E_1, E_2 are Borel sets with $E_1 \subseteq E_2$, then $C(E_1) \leq C(E_2)$.

Exercise 7.4: If E is a Borel set, then $C(E) = \sup C(F)$ taken over all compact $F \subseteq E$.

Proposition (Choquet): If E is a Borel set, then $C(E) = \inf C(\mathcal{O})$ taken over all open sets $\mathcal{O} \geq E$.

This last result is deep, and we shall make no use of it.

Exercise 7.5: If E_1, E_2 are Borel sets, then $C(E_1 \cup E_2) \leq C(E_1) + C(E_2)$.

We thus have that capacity, as a set function, shares several basic properties with Borel measures: capacity is monotone, subadditive, and regular (the last in virtue of Exercise 7.4 and Choquet's proposition).

However, capacity differs from measures in the following striking way: Let E be a compact set. Let $\bar{\mu}, \gamma$ be associated to E as above. By Theorem 7.1a, $\bar{\mu}$ lies on ∂E.

Also, $U^{\bar{\mu}/\gamma} \leq 1$ everywhere. Hence $\frac{1}{\gamma} \bar{\mu}(\partial E) \leq C(\partial E)$.

But $C(E) = \frac{1}{\gamma} \bar{\mu}(E)$, by Theorem 7.1c, and $\bar{\mu}(E) = \bar{\mu}(\partial E)$. Hence $C(E) \leq C(\partial E)$ and so equality holds. Thus:

Proposition: The capacity of a compact set equals the capacity of its boundary.

If E has non-empty interior, $C(\mathring{E}) > 0$ by Exercise 7.2. But E is the disjoint union of $\mathring{E}$ and ∂E. So capacity fails to be additive on disjoint sets.

8. Maximum Principle for Potentials

In this section we shall prove Theorem 7.4 which states that for a measure μ of compact support the value $U^\mu(x)$ of the potential of μ is dominated by the supremum of U^μ over supp μ, for every x in $\mathbb{R}^n$.

We shall deduce this result from the following:

Theorem 8.1: (Continuity Theorem) *Let* ν *be a measure with compact support* E. *Suppose* U^ν *is continuous when restricted to* E. *Then* U^ν *is continuous everywhere in* $\mathbb{R}^n$.

Let μ be a measure with compact support S such that $U^\mu \leq M$ on S. For each subset E of S denote by μ_E the restriction of μ to E.

Lemma 8.2: *Fix* $\epsilon > 0$. $\exists$ *compact set* $E \subseteq S$ *such that*

$$\mu(S\setminus E) < \epsilon, \tag{8.1}$$

$$U^{\mu_E} \text{ is continuous on } E. \tag{8.2}$$

Assume for the moment that both Theorem 8.1 and Lemma 8.2 are true. Choose E as in the Lemma and fix $x \in \mathbb{R}^3\setminus S$. Put d = distance (x,S). Set $\mu' = \mu_E$.

$$U^\mu(x) = (U^\mu - U^{\mu'})(x) + U^{\mu'}(x). \tag{8.3}$$

The first term on the right, using (8.1),

$$= \int_{S\setminus E} \frac{d\mu(y)}{|x-y|} \leq \frac{\mu(S\setminus E)}{d} \leq \frac{\epsilon}{d} .$$

Since $\mu' \leq \mu$, $U^{\mu'} \leq U^\mu \leq M$ on S. By (8.2) $U^{\mu'}$ is continuous on E, and supp $\mu' \subseteq E$. By Theorem 8.1, $U^{\mu'}$ is then continuous on $\mathbb{R}^3$. It follows by the classical maximum principle for harmonic functions, that

$$U^{\mu'}(x) \leq \max_S U^{\mu'} \leq M.$$

(8.3) now gives $U^{\mu}(x) \leq \frac{\epsilon}{d} + M$.

Since ϵ is arbitrary, we conclude $U^{\mu}(x) \leq M$, and since x was arbitrary in $\mathbb{R}^3 \setminus S$, we have proved Theorem 7.4.

Proof of Lemma 8.2: U^{μ} is a measurable function. By Lusin's theorem, $\exists$ a compact subset E of S such that $U^{\mu}|_E$ is continuous and $\mu(S \setminus E) < \epsilon$. We must prove (8.2).

Put $\mu' = \mu|_E$. Fix $\{x_n\} \in E$ with $x_n \to x_0$. We have to show that

$$\lim_{n \to \infty} U^{\mu'}(x_n) = U^{\mu'}(x_0).$$

By Lemma 7.5 applied to μ', $\underline{\lim}\, U^{\mu'}(x_n) \geq U^{\mu'}(x_0)$. It thus remains to show that $\overline{\lim}\, U^{\mu'}(x_n) \leq U^{\mu'}(x_0)$.

Put $\mu'' = \mu - \mu'$. Fix $\delta > 0$. μ'' is a measure of compact support, so by Lemma 7.5 applied to μ'',

$$U^{\mu''}(x_n) > U^{\mu''}(x_0) - \delta \quad \text{for large } n.$$

Also

$$U^{\mu}(x_n) < U^{\mu}(x_0) + \delta \quad \text{for large } n,$$

since U^{μ} is continuous on E. Hence

$$U^{\mu}(x_n) - U^{\mu''}(x_n) < U^{\mu}(x_0) - U^{\mu''}(x_0) + 2\delta,$$

or

$$U^{\mu'}(x_n) < U^{\mu'}(x_0) + 2\delta \quad \text{for large } n.$$

Hence $\overline{\lim}\, U^{\mu'}(x_n) \leq U^{\mu'}(x_0)$. The proof of Lemma 8.2 is complete.

For the proof of Theorem 8.1 we need the following characterization of continuity for potentials.

Lemma 8.3: Given a measure μ on $\mathbb{R}^3$ of compact support and given a compact set $K \subset \mathbb{R}^3$. In order that the restriction of U^{μ} to K is continuous it is necessary and sufficient that for each $\epsilon > 0$ $\exists \delta > 0$ satisfying

$$\int_{|x-y|<\delta} \frac{d\mu(y)}{|x-y|} < \epsilon \tag{8.4}$$

for every $x \in K$.

Proof: Assume for each $\epsilon > 0$, $\exists \delta > 0$ satisfying (8.4) for each $x \in K$.

As earlier, set

$$K_j(t) = \begin{cases} \frac{1}{t}, & t \geq \frac{1}{j} \\ j, & 0 \leq t < \frac{1}{j} \end{cases}$$

Fix ϵ and choose δ so that

$$\int_{|x-y|<\delta} \frac{d\mu(y)}{|x-y|} < \epsilon, \quad \text{all } x \in K.$$

Then

$$\int_{|x-y|\geq\delta} \frac{d\mu(y)}{|x-y|} > U^{\mu}(x) - \epsilon, \quad \text{all } x \in K.$$

For j with $\frac{1}{j} < \delta$, then, for all $x \in K$

$$\int K_j(|x-y|)d\mu(y) \geq \int_{|x-y|\geq 1/j} \frac{d\mu(y)}{|x-y|} > U^\mu(x) - \epsilon.$$

Also the left-side $\leq U^\mu(x)$ for all x. Hence, if $F_j(x) = \int K_j(|x-y|)d\mu(y)$, then $F_j \to U^\mu$ uniformly on K as $j \to \infty$. Since also F_j is continuous for each j, we conclude that U^μ is continuous and K, as desired.

Conversely, assume U^μ is continuous on K. Suppose (8.4) is false.

Then $\exists \epsilon > 0$ such that for each n we can find $x_n \in K$ with

$$\int_{|x_n-y|<1/n} \frac{d\mu(y)}{|x_n-y|} \geq \epsilon. \tag{8.5}$$

A subsequence of $\{x_n\}$, again denoted $\{x_n\}$, converges to some $\bar{x} \in K$.

For each $r > 0$, let

$$U_r(x) = \int_{|y-\bar{x}|\geq r} \frac{d\mu(y)}{|x-y|}.$$

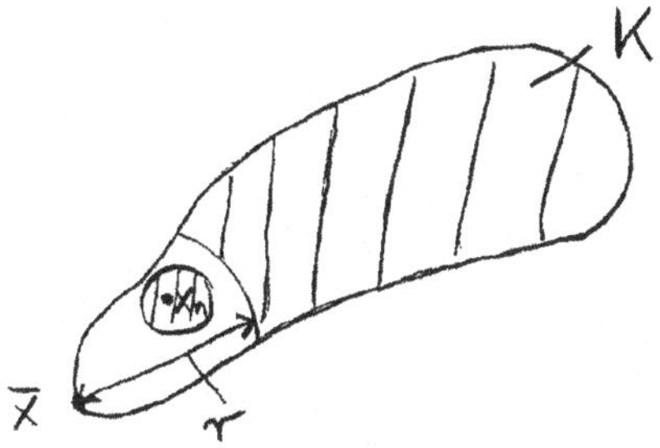

Then U_r is continuous on K at $\bar{x}$.

Put

$$V_r = U^\mu - U_r.$$

Since U^μ is continuous on K, V_r is continuous on K at $\bar{x}$. Now fix r.

For large n, the ball

$$\{y|\ |x_n - y| < \tfrac{1}{n}\} \subset \{y|\,|\bar{x}-y| < r\},$$

so

$$V_r(x_n) = \int_{|y-\bar{x}|<r} \frac{d\mu(y)}{|x_n-y|} \geq \int_{|y-x_n|<1/n} \frac{d\mu(y)}{|x_n-y|} \geq \epsilon.$$

By continuity, $V_r(\bar{x}) \geq \epsilon$. But since $U^\mu(\bar{x}) < \infty$, $U_r(\bar{x}) \to U^\mu(\bar{x})$ as $r \to 0$, or $V_r(\bar{x}) \to 0$. This is a contradiction. So (8.4) is true, and Lemma 8.3 is proved.

<u>Proof of Theorem 8.1</u>: We are given a measure ν of compact support E such that U^ν is continuous on E, and must prove U^ν continuous on $\mathbb{R}^3$

Fix a compact set K. We shall show that U^ν is continuous on K by verifying (8.4) for ν and K, and applying Lemma 8.3. Fix $\epsilon > 0$. The Lemma gives $\delta > 0$ such that

(8.6) $$\int_{|x-y|<\delta} \frac{d\nu(y)}{|x-y|} < \frac{\epsilon}{2}$$

for every $x \in E$.

Fix $x_o \in K$

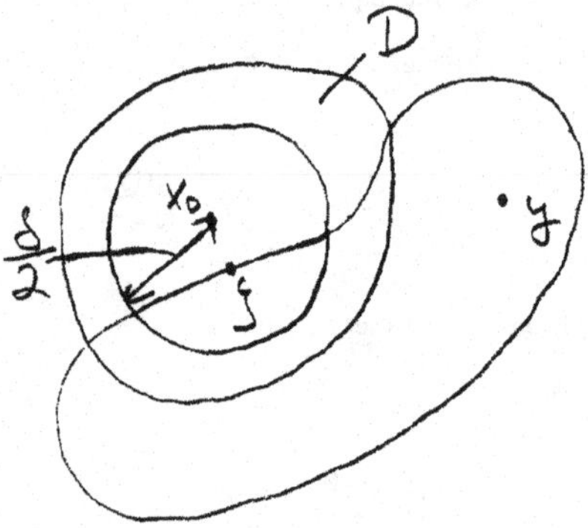

<u>Assertion 1</u>: $$\int_{|x_o-y|<\delta/2} \frac{d\nu(y)}{|x_o-y|} < \epsilon.$$

For if the distance from x_o to E $\geq \frac{\delta}{2}$, then $\{y \mid |x_o - y| < \frac{\delta}{2}\}$ lies outside E, so the integral $= 0$.

If the distance $< \frac{\delta}{2}$, then $|x_o - \xi| < \frac{\delta}{2}$, where ξ is the nearest point to x_o in E. Let D be the ball $\{y \mid |y-\xi| < \delta\}$. The ball $\{y \mid |x_o - y| < \frac{\delta}{2}\}$ is contained in D.

Fix $y \in E$. Then

$$|\xi-y| \leq |\xi-x_o| + |x_o - y|,$$

whence

$$\frac{|\zeta-y|}{|x_o-y|} \leq \frac{|\zeta-x_o|}{|y-x_o|} + 1 \leq 2,$$

by choice of ζ. Hence

$$\int_{|x_o-y|<\delta/2} \frac{d\nu(y)}{|x_o-y|} \leq \int_D \frac{d\nu(y)}{|x_o-y|}$$

$$= \int_D \frac{|\zeta-y|}{|x_o-y|} \frac{d\nu(y)}{|\zeta-y|} \leq 2 \int_D \frac{d\nu(y)}{|\zeta-y|},$$

by (8.7). Since $\zeta \in E$, the integral on the right $< \frac{\epsilon}{2}$ by (8.6). Thus

$$\int_{|x_o-y|<\delta/2} \frac{d\nu(y)}{|x_o-y|} \leq 2 \cdot \frac{\epsilon}{2} = \epsilon.$$

Assertion 1 thus holds.

By Lemma 8.3, Assertion 1 yields that U^{ν} is continuous on K. But K was an arbitrary compact set. So the proof of Theorem 8.1 is complete.

9. Uniqueness of the Equilibrium Distribution

We shall use the following:

Notation: A statement concerning points in $\mathbb{R}^n$ is true p.p. if the set of points where it fails to hold has capacity 0.

In Theorem 7.1 we showed the following: given a compact set E in $\mathbb{R}^n$ $\exists$ a probability measure $\bar{\mu}$ on E and a constant γ such that

(9.1) $$U^{\bar{\mu}} = \gamma \text{ p.p. on } E \text{ and } U^{\bar{\mu}} \leq \gamma \text{ everywhere}$$

We shall now show that there is precisely one such measure.

Theorem 9.1: Fix a compact set E in $\mathbb{R}^n$. Let $\bar{\mu}, \gamma$ satisfy (9.1) and let μ' be a probability measure on E and γ' a constant such that

(9.2) $$U^{\mu'} = \gamma' \text{ p.p. on } E \text{ and } U^{\mu'} \leq \gamma' \text{ everywhere.}$$

Then $\mu' = \bar{\mu}$.

Note: It follows at once that $\exists$ a unique probability measure on E which minimizes energy, since Theorem 7.1 showed that if a measure on E minimizes energy then (9.1) holds.

We need the following fact: if ν is a signed measure, i.e., $\nu = \alpha - \beta$, where α, β are measures, and if

$$[\nu,\nu] = \int \frac{d\nu(x)d\nu(y)}{|x-y|} = 0,$$

then $\nu = 0$. By Theorem 6.2, this result is true for signed measures of the form:

$\nu = \rho dx$, $\rho \in C_0^2(\mathbb{R}^3)$, for the theorem gives that $I(\nu) = 0$ implies grad $U^\nu = 0$ everywhere, and so U^ν is constant in $\mathbb{R}^3$, whence U^ν is identically 0. But $\nu = \mu_1 - \mu_2$, where μ_1, μ_2 are measures, so $U^{\mu_1} = U^{\mu_2}$ everywhere. By Corollary 1 of Theorem 4.1, this implies $\mu_1 = \mu_2$, or $\nu = 0$ as desired.

We need the general result:

Theorem 9.2: *Let* $\sigma = \sigma_1 - \sigma_2$, *where* σ_1, σ_2 *are measures of compact support on* $\mathbb{R}^n$. *Assume*

(9.3) $$I(\sigma_1 + \sigma_2) < \infty.$$

Then

(9.4) $$I(\sigma) = [\sigma,\sigma] \geq 0,$$

and if

(9.5) $$I(\sigma) = 0, \text{ then } \sigma = 0.$$

Assume Theorem 9.2 is proved.

Proof of Theorem 9.1:

$$\int U^{\bar{\mu}} d\mu' = \int U^{\mu'} d\bar{\mu}, \quad \text{or} \quad \gamma = \gamma'.$$

$$[\bar{\mu}-\mu', \bar{\mu}-\mu'] = I(\bar{\mu}) + I(\mu') - 2[\bar{\mu},\mu'].$$

$$I(\bar{\mu}) = \gamma, \quad I(\mu') = \gamma' = \gamma.$$

So

$$[\bar{\mu}-\mu', \bar{\mu}-\mu'] = \gamma + \gamma - 2\gamma = 0.$$

Also,

$$I(\bar{\mu}+\mu') = \int (U^{\bar{\mu}} + U^{\mu'})(d\bar{\mu}+d\mu') < \infty,$$

since $U^{\bar{\mu}}, U^{\mu'}$ are bounded functions. Thus $\bar{\mu} - \mu'$ satisfies condition (9.3). Since $I(\bar{\mu}-\mu') = 0$, $\bar{\mu} - \mu' = 0$ by Theorem 9.2. So Theorem 9.1 holds.

We need

Exercise 9.1: There is a constant $\beta > 0$ so that for all $x,y \in \mathbb{R}^3$ we have

$$\frac{\beta}{|x-y|} = \int \frac{d\xi}{|x-\xi|^2 \cdot |y-\xi|^2}. \tag{9.6}$$

Proof of Theorem 9.2: $\sigma = \sigma_1 - \sigma_2$. Let $|\sigma|$ denote the total variation measure of σ. We claim

$$\int d\sigma(x) \int d\sigma(y) \int \frac{d\xi}{|x-\xi|^2 |y-\xi|^2} = \int d\xi \int \frac{d\sigma(x)}{|x-\xi|^2} \int \frac{d\sigma(y)}{|y-\xi|^2}. \tag{9.7}$$

Fubini's theorem yields (9.7) provided

$$\int d|\sigma|(x) \int d|\sigma|(y) \int \frac{d\xi}{|x-\xi|^2 |y-\xi|^2} < \infty. \tag{9.8}$$

But the left side =

$$\beta \int d|\sigma|(x) \int \frac{d|\sigma|(y)}{|x-y|} = \beta I(|\sigma|),$$

by Exercise 9.1. Also $|\sigma| \le \sigma_1 + \sigma_2$, so $I(|\sigma|) \le I(\sigma_1 + \sigma_2) < \infty$ by (9.3). So (9.8) holds, and so (9.7) follows. Then

$$\beta \int \frac{d\sigma(x) d\sigma(y)}{|x-y|} = \int d\xi \left\{ \int \frac{d\sigma(x)}{|x-\xi|^2} \right\}^2, \tag{9.9}$$

by (9.7), where we again used Exercise 9.1.

(9.9) gives assertion (9.4). If $I(\sigma) = 0$, then (9.9) yields

(9.10) $$\int \frac{d\sigma(x)}{|x-\xi|^2} = 0 \quad \text{for a.a. } \xi \text{ in } \mathbb{R}^3.$$

(9.10) implies

$$\int d\xi \, \frac{1}{|y-\xi|^2} \int \frac{d\sigma(x)}{|x-\xi|^2} = 0$$

for all y $\mathbb{R}^3$, and so

$$\int d\sigma(x) \int \frac{d\xi}{|x-\xi|^2 |y-\xi|^2} = 0 \text{ for a.a. } y,$$

or

$$\int \frac{d\sigma(x)}{|x-y|} = 0 \quad \text{for a.a. } y,$$

or

$$\int \frac{d\sigma_1(x)}{|x-y|} = \int \frac{d\sigma_2(x)}{|x-y|} \text{ a.e. .}$$

By Corollary 1 to Theorem 4.1 it follows that $\sigma_1 = \sigma_2$, whence $\sigma = 0$. So (9.5) holds, and Theorem 9.2 is proved.

Note: In view of Theorem 9.1 we can speak of the equilibrium distribution for a given set E.

Exercise 9.2: Let E be a compact set in $\mathbb{R}^n$. Let $\bar{\mu}$ be its equilibrium distribution and γ the number such that $U^{\bar{\mu}} = \gamma$ p.p. on E.

Then for every probability measure σ on E we have

(9.11) $$\inf_{x \in E} U^{\sigma}(x) \leq \gamma \leq \sup_{x \in E} U^{\sigma}(x).$$

10. The Cone Condition

Theorem 7.1 left us with the following problem: let E be a compact set, $\bar{\mu}$ its equilibrium distribution with $U^{\bar{\mu}} = \gamma$ p.p. on E, and $\tilde{E}$ the exceptional set, i.e.,

$$\tilde{E} = \{x \in E \mid U^{\bar{\mu}}(x) < \gamma\}.$$

When does a given point of E lie in $E \setminus \tilde{E}$?

Theorem 10.1: Let E be a compact set in $\mathbb{R}^3$ and fix $x_0 \in E$. Assume $\exists K > 0$ such that for all sufficiently small $r > 0$

(10.1) $$C(E \cap \{x \mid |x-x_0| \leq r\}) > K \cdot r.$$

Then $x_0 \in E \setminus \tilde{E}$, i.e., $U^{\bar{\mu}}(x_0) = \gamma$.

Proof: Observe that if E_a is a translate of E, i.e., $E_a = \{x + a \mid x \in E\}$ then

(10.2) $$C(E) = C(E_a),$$

and

(10.3) The equilibrium potential of E_a is a translate of the equilibrium potential of E.

Hence in proving Theorem 10.1 we may without loss of generality suppose that $x_0 = 0$. Put

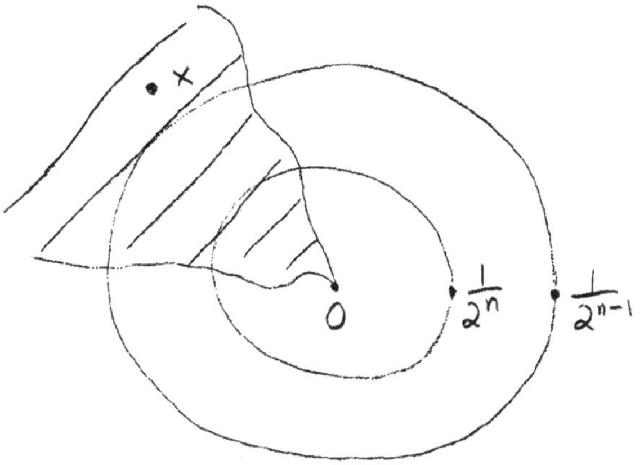

$E_n = E \cap \{x \mid |x| \leq \frac{1}{2^n}\}$.

For large n, (10.1) gives

$$C(E_n) > \frac{K}{2^n} .$$

We consider only such n.

For each n let σ_n be the equilibrium distribution for E_n and $\gamma_n = I(\sigma_n)$. Then $U^{\sigma_n} = \gamma_n$ p.p. on E_n, and $C(E_n) = \frac{1}{\gamma_n}$.

Fix $x \in \mathbb{R}^3$. We claim that for all n:

$$U^{\sigma_n}(x) \leq \frac{2}{K} \cdot \frac{1}{|x|} . \tag{10.4}$$

Case 1: $|x| \leq \frac{1}{2^n}$. We have

$$U^{\sigma_n}(x) \leq \gamma_n = \frac{1}{C(E_n)} \leq \frac{1}{K} \cdot 2^n,$$

by (10.1). Also $\frac{1}{|x|} \geq 2^n$, so (10.4) holds.

Case 2: $|x| \geq \frac{1}{2^{n-1}}$.

For $|y| \leq \frac{1}{2^n}$, we have

$$|x-y| \geq |x| - \frac{1}{2^n} \geq |x| - \frac{|x|}{2} = \frac{|x|}{2} ,$$

so

$$U^{\sigma_n}(x) = \int \frac{d\sigma_n(y)}{|x-y|} \le \frac{2}{|x|} \int d\sigma_n(y) = \frac{2}{|x|} .$$

Without loss of generality $K < \frac{1}{2}$. Then $\frac{2}{|x|} < \frac{1}{K}\frac{1}{|x|}$, i.e., (10.4).

<u>Case 3</u>: $\frac{1}{2^n} < |x| < \frac{1}{2^{n-1}}$.

By Theorem 7.4,

$$U^{\sigma_n}(x) \le \sup_{E_n} U^{\sigma_n} = \gamma_n \le \frac{2^n}{K} .$$

Also $\frac{1}{|x|} > 2^{n-1}$, so $U^{\sigma_n}(x) \le \frac{2}{K}\frac{1}{|x|}$, i.e., (10.4) once more, and so (10.4) holds for all x, as claimed. Next we claim

(10.5) $$\int \frac{d\sigma_n(y)}{|x-y|} \to \frac{1}{|x|} \quad \text{as} \quad n \to \infty,$$

for $x \ne 0$.

For, with x fixed, $f(y) = \frac{1}{|x-y|}$ is continuous on E_n for large n, so

$$\int \frac{d\sigma_n(y)}{|x-y|} - \frac{1}{|x|} = \int (f(y) - f(0))d\sigma_n(y) \to 0,$$

since $\operatorname{supp} \sigma_n \subseteq \{y \mid |y| \le \frac{1}{2^n}\}$.

<u>Assertion</u>: If μ is a measure of compact support such that $U^\mu(0) < \infty$, then

(10.6) $$\int U^\mu(x)d\sigma_n(x) \to U^\mu(0).$$

For

$$\int U^{\mu}(x)d\sigma_n(x) = \int \left\{ \int \frac{d\mu(y)}{|x-y|} \right\} d\sigma_n(x)$$

$$= \int d\mu(y) \left\{ \int \frac{d\sigma_n(x)}{|x-y|} \right\} \to \int \frac{1}{|y|} d\mu(y) = U^{\mu}(0)$$

by (10.5) and the fact that in view of (10.4) and the assumption that $U^{\mu}(0) < \infty$, the convergence is dominated.

So (10.6) holds.

Let now $\bar{\mu}$, γ, E, $\tilde{E}$ be as in the statement of the theorem. Fix n.

$$\int_E U^{\bar{\mu}} d\sigma_n = \int_{E\setminus\tilde{E}} U^{\bar{\mu}} d\sigma_n + \int_{\tilde{E}} U^{\bar{\mu}} d\sigma_n .$$

$C(\tilde{E}) = 0$, so $\sigma_n(\tilde{E}) = 0$ by Lemma 7.3, so the second integral on the right vanishes. $U^{\bar{\mu}} = \gamma$ on $E\setminus\tilde{E}$ and $\sigma_n(E\setminus\tilde{E}) = 1$, so the first integral on the right $= \gamma$, and so

$$\int_E U^{\bar{\mu}} d\sigma_n = \gamma \quad \text{for all } n.$$

By (10.6) this implies $U^{\bar{\mu}}(0) = \gamma$. q.e.d.

In order to get practical use out of the result just proved, we need geometric conditions assuming that (10.1) holds.

<u>Definition 10.1</u>: Fix $x \in \mathbb{R}^n$. By a <u>cone</u> K_x with vertex at x we mean a set obtained as follows:

Fix a ray L emanating from x and fix $\alpha > 0$. K_x is the union of all rays L^1 starting at x which make an angle $\leq \alpha$ with L.

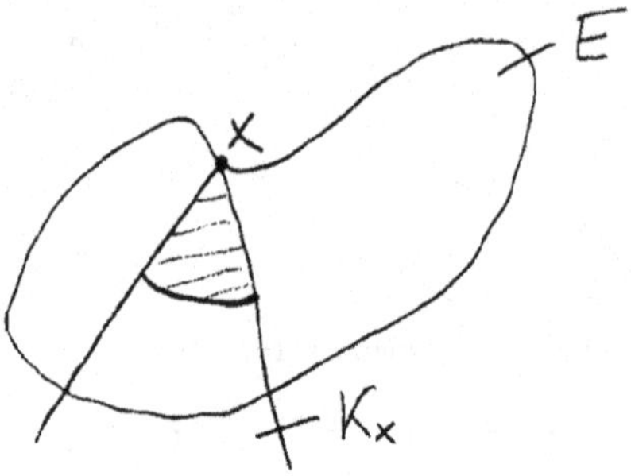

Definition 10.2: Let E be a closed set in $\mathbb{R}^n$. Fix $x \in E$. E satisfies the cone condition at x if $\exists$ a cone K_x with vertex at x such that for some r_o E contains

$$K_x \cap \{y \mid |y-x| \leq r_o\}.$$

Theorem 10.2: Let E be a compact set in $\mathbb{R}^n$, $\bar{\mu}$ its equilibrium distribution with $U^{\bar{\mu}} = \gamma$ p.p. on E. Fix $x_o \in E$ such that E satisfies the cone condition at x_o. Then $U^{\bar{\mu}}(x_o) = \gamma$.

Proof: Without loss of generality $x_o = 0$. Fix r_o and a cone K with vertex at 0 such that E contains $K \cap \{x \mid |x| \leq r_o\}$. Fix $r \leq r_o$ and set $K^{(r)} = K \cap \{x \mid |x| \leq r\}$. We can choose finitely many congruent copies of $K^{(r)}, K_1^{(r)}, \ldots, K_s^{(r)}$ such that

$$\{x \mid |x| \leq r\} \subset \bigcup_{i=1}^{s} K_i^{(r)}.$$

Exercise 10.1: Let F be a compact set in $\mathbb{R}^n$ and F^1 a congruent set, i.e., $\exists$ Euclidean motion T such that $T(F) = F^1$. Then $C(F) = C(F^1)$.

By the subadditive property of capacity (recall Exercise 7.5),

$$C\{x \mid |x| \leq r\} \leq \sum_{i=1}^{s} C(K_i^{(r)}).$$

The right side $= sC(K^{(r)})$, by Exercise 10.1, while the left side $= r$, as was seen in Section 5. Hence

$$\frac{r}{s} \le C(K^{(r)}) \le C(E \cap \{x|\ |x| \le r\}),$$

since $K^{(r)} \subseteq E$. Hence (10.1) is satisfied, and Theorem 10.1 yields the assertion.

Corollary: Let Ω be a bounded region in $\mathbb{R}^n$ whose boundary is a smooth manifold. Let $E = \overline{\Omega}$. Then $\tilde{E}$ is empty, i.e., $U^{\overline{\mu}} = \gamma$ everywhere on E.

Proof: The smoothness of the boundary evidently implies that E satisfies the cone condition at each point of ∂E.

Note: Condition (10.1) is merely sufficient in order that $U^{\overline{\mu}}(x_0) = \gamma$, but is not necessary. A necessary and sufficient condition was given by Norbert Wiener in 1924. It is rather more complicated than (10.1).

Exercise 10.2: Construct a compact set X in $\mathbb{R}^3$ with equilibrium distribution $\overline{\mu}$ such that $U^{\overline{\mu}} = \gamma$ on X p.p., but $\exists\, x_0 \in X$, x_0 not an isolated point of X, with $U^{\overline{\mu}}(x_0) < \gamma$.

11. Singularities of Bounded Harmonic Functions

Let Ω be an open set in $\mathbb{R}^n$ and E a compact subset of Ω.

Problem: Find conditions on E in order that every bounded harmonic function on $\Omega\setminus E$ extends to all of Ω, i.e., in order that whenever u is harmonic in $\Omega\setminus E$ and $|u| < M$ for some M, then $\exists\, \tilde{u}$ harmonic in Ω with $\tilde{u} = u$ in $\Omega\setminus E$.

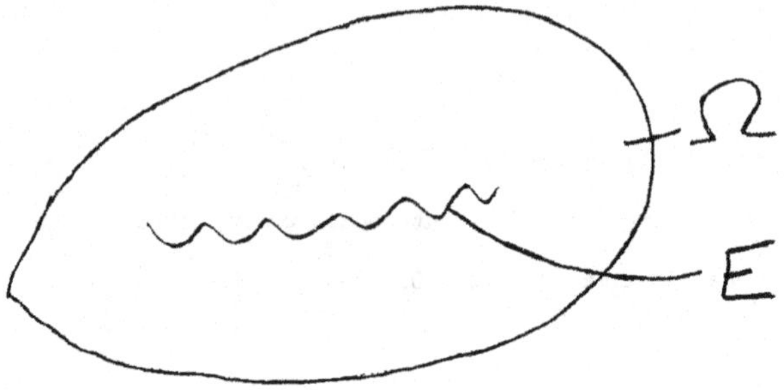

The best known example of this situation is when Ω is a disk $\{z \mid |z-z_0| < r\}$ in $\mathbb{R}^2$ and $E = \{z_0\}$. Let u be harmonic and bounded in $\Omega\setminus E = \{z \mid 0 < |z-z_0| < r\}$. The multiple-valued conjugate v of u in $\Omega\setminus E$ has a certain period on any cycle round z_0. A suitable multiple tu then has the conjugate tv with period 2π. Hence

$$f(z) = e^{tu+itv}$$

is a single-valued analytic function in $\Omega\setminus E$, and $|f|$ is bounded, since $|f| = e^{tu}$. Hence by Riemann's theorem, z_0 is a removable singularity for f, so $\exists\, \tilde{f}$ analytic in $|z-z_0| < r$ with $\tilde{f} = f$ for $z \neq z_0$. Also

$$\tilde{f}(z_0) \neq 0, \quad \text{for else}$$

$\lim_{z \to z_0} e^{tu(z)} = 0$ which contradicts the fact that u is bounded. Hence $\log \tilde{f}$ is single-valued analytic in some neighborhood of z_0. Put

$$\tilde{u} = \frac{1}{t} \operatorname{Re}(\log \tilde{f}).$$

Then $\tilde{u}$ is harmonic in a neighborhood of z_0, and

$$\tilde{u} = \frac{1}{t} \operatorname{Re}(\log f) = \frac{1}{t} \log |f| = u,$$

in a deleted neighborhood of z_0.

To attack our problem in $\mathbb{R}^n$, $n > 2$, we need the following fundamental result about harmonic functions, which will be proved in a later section.

Proposition 11.1: Let W be a bounded open subset of $\mathbb{R}^n$. Assume ∂W is the union of finitely many smooth compact manifolds. Then to each continuous function f defined on ∂W, there corresponds a harmonic function F in W which assumes f continuously as its boundary values.

We next look at the analogue of the punctured disk in $\mathbb{R}^3$.

Assertion: Given u harmonic in the punctured ball: $0 < |x| < R$ in $\mathbb{R}^3$, with $|u| \leq M$. Then u extends to the full ball.

For fix $R^1 < R$. Restricted to $|x| = R^1$, u is continuous. By Proposition 11.1, we can find U harmonic in $|x| \leq R^1$ with $U = u$ on $|x| = R^1$.

To prove our assertion, it is enough to show that $U = u$ in $0 < |x| < R^1$. Fix n and let W_n be the region:

$$W_n = \{x \mid \tfrac{1}{n} < |x| < R^1\}.$$

Note that by the maximum principle,

$$|U| \leq M \quad \text{in} \quad |x| \leq R^1. \tag{11.1}$$

We compare the two functions $u - U$ and $\frac{2M}{n} \cdot \frac{1}{|x|}$. Both are harmonic in W_n. On $|x| = R^1$,

$$u - U = 0 \leq \frac{2M}{n} \cdot \frac{1}{|x|} .$$

On $|x| = \frac{1}{n}$,

$$u - U \leq 2M \leq \frac{2M}{n} \cdot \frac{1}{|x|} .$$

Hence by the maximum principle applied to $u - U$ in W_n,

$$u(x) - U(x) \leq \frac{2M}{n} \cdot \frac{1}{|x|}, \quad x \text{ in } W_n. \tag{11.2}$$

Fix x, $0 < |x| < R^1$. For all large n, $x \in W_n$, so (11.2) holds and yields, as $n \to \infty$

$$u(x) - U(x) \leq 0.$$

The same reasoning applied to $U - u$, yields the reverse inequality, whence

$$u(x) = U(x).$$

So the assertion is proved.

We now want to imitate the preceding argument in the general case. Let Ω be an open set in $\mathbb{R}^3$, E a compact subset of Ω, u a function harmonic in $\Omega \setminus E$ with $|u| \leq M$.

Since we can replace Ω by a slightly smaller region, we may assume without loss of generality that $\partial\Omega$ is a finite union of disjoint smooth closed surfaces and that u is continuous on $\bar{\Omega}$.

For each n, choose a smoothly bounded open set Ω_n with $E \subset \Omega_n$, such that

(11.3) $$\bar{\Omega}_{n+1} \subset \Omega_n \quad \text{and} \quad \bigcap_{n=1}^{\infty} \Omega_n = E.$$

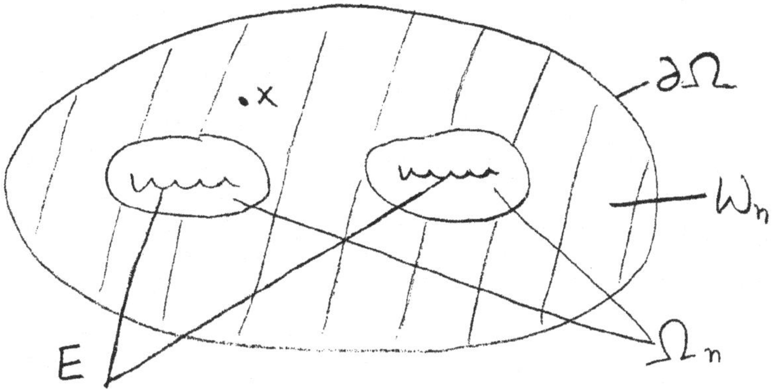

Let $W_n = \Omega \backslash \bar{\Omega}_n$. We now endeavor to obtain functions φ_n in W_n so that

(11.4) φ_n is harmonic in W_n and continuous on $\partial W_n = \partial\Omega_n \cup \partial\Omega$.

(11.5) $\varphi_n = 1$ on $\partial\Omega_n$.

(11.6) $\varphi_n \geq 0$ on $\partial\Omega$.

(11.7) For each x in $\Omega \backslash E$, $\lim\limits_{n \to \infty} \varphi_n(x) = 0$.

Assuming the φ_n to have been constructed, we proceed as follows:

Using Proposition 11.1, we choose U harmonic in Ω, continuous on $\bar{\Omega}$ and $U = u$ on $\partial\Omega$. Then

(11.8) $U - u = 0$ on $\partial\Omega$, $U - u \leq 2M$ on $\partial\Omega_n$.

(11.9) $2M\varphi_n \geq 0$ on $\partial\Omega$, $2M\varphi_n = 2M$ on $\partial\Omega_n$.

Hence $U - u \leq 2M\varphi_n$ in W_n.

Fix $x \in \Omega\backslash E$. Because of (11.3), $x \notin \overline{\Omega}_n$ for large n, so $x \in W_n$ for large n. (11.7) then gives

$$U(x) - u(x) \leq 0.$$

Similarly $u(x) - U(x) \leq 0$. Hence $u = U$ in $\Omega\backslash E$.

We must, therefore, find conditions on E which allow us to construct the φ_n. Now a natural candidate for φ_n is obtained from the equilibrium distribution for $\overline{\Omega}_n$.

Let μ_n be that measure. Then $U^{\mu_n} = \gamma_n$ on $\overline{\Omega}_n$, the exceptional set here being empty since $\overline{\Omega}_n$ satisfies the cone condition at each point, being a smoothly bounded set. Also

$$C(\overline{\Omega}_n) = \frac{1}{\gamma_n}.$$

Put $\varphi_n = \frac{1}{\gamma_n} U^{\mu_n}$. Then φ_n satisfies (11.4), (11.5), (11.6) because of the choice of μ_n. It remains to check (11.7).

$$\varphi_n(x) = \frac{1}{\gamma_n} \int \frac{d\mu_n(y)}{|x-y|}.$$

If $x \notin E$, $\exists d > 0$ (d depending on x) such that $|x-y| \geq d$ for all $y \in \overline{\Omega}_n$, provided n is large enough. Hence

$$\varphi_n(x) \leq \frac{1}{\gamma_n} \cdot \frac{1}{d}.$$

To have (11.7) we thus need

$$C(\overline{\Omega}_n) = \frac{1}{\gamma_n} \to 0 \quad \text{as} \quad n \to \infty. \tag{11.10}$$

In view of (11.3) it seems likely that (11.10) holds provided $C(E) = 0$. This is in fact true, as follows from the next lemma.

Lemma 11.1: *Let* K *be a compact set in* R^k *and let* $\{U_n\}$ *be a sequence of bounded open neighborhoods of* K *such that* $\overline{U}_{n+1} \subseteq U_n$ *and* $\bigcap_{n=1}^{\infty} U_n = K$.

Then $C(K) = \lim_{n \to \infty} C(U_n)$.

Proof: Since $U_{n+1} \subseteq U_n$, the sequence $\{C(U_n)\}$ is monotone and so the limit exists. Since $K \subset U_n$ for each n,

$$C(K) \leq \lim_{n \to \infty} C(U_n).$$

We must show the reverse inequality.

Let L be a compact neighborhood of K such that $U_n \subseteq L$ for all n. For each n choose a measure μ_n with $\operatorname{supp} \mu_n \subset U_n$, $U^{\mu_n} \leq 1$ on U_n and

$$\mu_n(U_n) \geq C(U_n) - \frac{1}{n}. \tag{11.11}$$

The definition of capacity allows us to do this.

Then the total mass of $\mu_n \leq C(U_n) \leq C(L)$, for all n. Hence $\exists$ a weakly convergent subsequence of $\{\mu_n\}$, again denoted $\{\mu_n\}$, on L, whose limit is a measure μ on L.

Fix n_o. For $n > n_o$,

$$\operatorname{supp} \mu_n \subseteq U_n \subset \overline{U}_{n_o}.$$

Hence $\operatorname{supp} \mu \subset \overline{U}_{n_o}$. Since n_o was arbitrary, $\operatorname{supp} \mu \subset K$.

$$\mu_n(U_n) = \int_L 1 d\mu_n \to \int_L 1 d\mu = \mu(K).$$

(11.11) thus gives

(11.12) $$\mu(K) \geq \lim_{n \to \infty} C(U_n).$$

We next claim

(11.13) $$\mu(K) \leq C(K).$$

(11.12) and (11.13) yield $C(K) \geq \lim_{n \to \infty} C(U_n)$, as desired.

It remains to verify (11.13). We need:

Exercise 11.1: Let $\{\sigma_n\}$ be a sequence of measures on a compact set E in $\mathbb{R}^k$ with σ_n converging weakly on E to a measure σ. Then for all $x \in E$,

$$U^{\sigma}(x) \leq \underline{\lim}_{n \to \infty} U^{\sigma_n}(x).$$

Now in our case, for $x \in K$ and each n, $U^{\mu_n}(x) \leq 1$. Hence by the Exercise, $U^{\mu}(x) \leq 1$. It follows that $\mu(K) \leq C(K)$, i.e., (11.13) holds. The proof of Lemma 11.1 is complete.

Thus (11.10) follows if $C(E) = 0$. Putting it all together we have: if $C(E) = 0$, then a function u bounded and harmonic on $\Omega \setminus E$ extends to all of Ω to be harmonic.

What about the converse? Assume $C(E) > 0$. By definition of capacity, $\exists\, \mu$ on E, $\mu \neq 0$, such that $U^{\mu} \leq 1$ on E. Then U^{μ} is harmonic on $\Omega \setminus E$ (in fact on $\mathbb{R}^3 \setminus E$) and, by the maximum principle for potentials, $U^{\mu} \leq 1$ everywhere.

We claim that U^{μ} cannot be extended to a function harmonic in Ω. For if so, $\exists$ a function V harmonic in all $\mathbb{R}^3$ with $V = U^{\mu}$ on $\mathbb{R}^3 \setminus E$.

Fix $\epsilon > 0$. On a large sphere, $|x| = R$, $V < \epsilon$. Hence $V < \epsilon$ in $|x| \leq R$. Hence $U^{\mu} < \epsilon$ in $(\mathbb{R}^3 \setminus E) \cap (|x| \leq R)$. Hence $U^{\mu} \leq 0$ in $(\mathbb{R}^3 \setminus E) \cap (|x| \leq R)$. But $U^{\mu} > 0$ everywhere, since $\mu \neq 0$. This contradiction proves our claim.

In sum, we have proved,

Theorem 11.2: *Let* Ω *be an open set in* $\mathbb{R}^3$ *and* E *a compact subset of* Ω. *In order that every function bounded and harmonic in* $\Omega \setminus E$ *admit an extension to all of* Ω, *it is necessary and sufficient that* $C(E) = 0$.

Note: The same theorem, with essentially the same proof, holds in $\mathbb{R}^n$, $n > 3$.

Exercise 11.2: Let E be a compact set in $\mathbb{R}^n$. If $\exists$ measure σ on E such that $U^\sigma = \infty$ at each point of E, then $C(E) = 0$.

Exercise 11.3: Let I be a linear segment in $\mathbb{R}^3$. Then $C(I) = 0$.

12. Green's Function

Earlier we considered the phenomenon of a grounded conducting surface surrounding a charged body. We saw that the surface acquired an induced charge, and the combined potential of the charge on the body and the induced charge on the surface equals zero on the surface.

Suppose now the body is replaced by a unit point charge at a point x_0 inside the surface Σ.

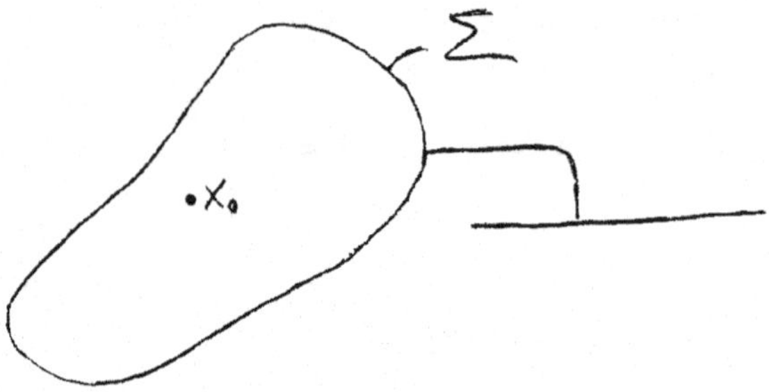

Again there results an induced charge on Σ, given by a set function $-\mu$ on Σ which is the negative of a measure μ. The combined potential U due to the point charge and the induced charge on Σ is given by:

$$U(x) = \frac{1}{|x-x_0|} - \int \frac{d\mu(\zeta)}{|x-\zeta|} .$$

U satisfies

(a) $U = 0$ on Σ.

(b) $U - \dfrac{1}{|x-x_0|}$ is harmonic inside Σ.

(b) says that U is harmonic in $\Omega\setminus\{x_0\}$, where Ω is the interior of Σ, and has at x_0 the singularity $\dfrac{1}{|x-x_0|}$.

The preceding reasoning, which leads to a function U satisfying (a) and (b) above, was made by George Green in his "Essay on the Application of Mathematical Analysis to the Theory of Electricity and Magnetism", Nottingham, 1828.

Definition 12.1: Let Ω be a region in $\mathbb{R}^3$ and x_0 a point in Ω. Let G_{x_0} be a function defined and continuous in $\bar{\Omega}\setminus\{x_0\}$ such that

(12.1) $$G_{x_0}(x) = 0 \quad \text{when} \quad x \in \partial\Omega.$$

(12.2) $$G_{x_0} \quad \text{is harmonic in} \quad \Omega\setminus\{x_0\}.$$

(12.3) $$G_{x_0} \quad \text{has at} \quad x_0 \quad \text{the singularity} \quad \frac{1}{|x-x_0|}.$$

Then G_{x_0} is called the Green's function of Ω with pole at x_0.

Note: While the existence of the Green's function, in a mathematical sense, remains to be proved, and we shall give one such proof in the next section, uniqueness is evident.

For if G and G' satisfy (12.1), (12.2) and (12.3), then $G - G'$ is harmonic in Ω and $=0$ on $\partial\Omega$, and so vanishes identically in Ω.

Let Ω be a smoothly bounded domain in $\mathbb{R}^3$, $x_0 \in \Omega$. Assume the Green's function $G = G_{x_0}$ exists for Ω. Assume in addition

(12.4) G has continuous first derivatives in the closed region $\bar{\Omega}$.

Theorem 12.1: Under the above hypothesis, for every function u harmonic in Ω and continuous in $\bar{\Omega}$, we have

(12.5) $$u(x_0) = -\frac{1}{4\pi}\int_{\partial\Omega} u \frac{\partial G}{\partial n}\, dS,$$

where $\frac{\partial}{\partial n}$ = exterior normal derivative.

Note: (12.5) allows us to calculate the value at x_o of a harmonic function in terms of its boundary values on $\partial\Omega$. That a formula of the type of (12.5) must exist follows from very general considerations: Let W be an arbitrary bounded domain in $\mathbb{R}^n$, without any hypotheses on ∂W, and fix $x_o \in W$. Let H be the vector space of functions in $C(\partial W)$ which extend to W to be harmonic. Call φ the linear functional on H given by

$$\varphi(f) = f(x_o), \quad \text{all } f \in H.$$

By the maximum principle,

$$(12.6) \qquad |\varphi(f)| \leq \max_{\partial W} |f|, \quad \text{all } f \in H.$$

Now H is a subspace of the Banach space $C(\partial W)$ of all continuous functions on ∂W, with

$$\|f\| = \max_{\partial W} |f|.$$

It follows from (12.6) and the Hahn-Banach theorem that φ admits an extension to a linear functional $\tilde{\varphi}$ on $C(\partial W)$, with $\|\tilde{\varphi}\| \leq 1$. By the Riesz representation theorem for linear functionals, $\exists$ a signed measure σ on ∂W with

$$(12.7) \qquad \tilde{\varphi}(f) = \int f d\sigma, \text{ all } f \in C(\partial W),$$

and

$$(12.8) \qquad \int d|\sigma| \leq 1.$$

Also

$$(12.9) \qquad 1 = \varphi(1) = \tilde{\varphi}(1) = \int d\sigma.$$

(12.8) and (12.9) imply that $\sigma > 0$, the proof of this implication being an exercise. We have proved:

Proposition: If Ω is a bounded domain in $\mathbb{R}^n$ and $x_0 \in \Omega$, then $\exists$ measure σ on ∂W (σ depending on x_0) such that for every function u harmonic in Ω and continuous in $\bar{\Omega}$ we have

$$(12.10) \qquad u(x_0) = \int_{\partial W} u(x)\,d\sigma(x).$$

Theorem 12.1 gives that, for a smoothly bounded domain Ω in $\mathbb{R}^3$ possessing a Green's function G which is smooth on $\bar{\Omega}$, the measure σ in (12.10) can be chosen to be

$$d\sigma = -\frac{1}{4\pi}\frac{\partial G}{\partial n}\,dS.$$

Proof of Theorem 12.1: Let Ω_ϵ be the region obtained from Ω by removing the ball $\{x|\ |x-x_0| \le \epsilon\}$.

$$\int_{\partial\Omega_\epsilon} \left(u\frac{\partial G}{\partial n} - G\frac{\partial u}{\partial n}\right)dS = \int_{\Omega_\epsilon} (u\Delta G - G\Delta u)\,dx.$$

The right-side vanishes since u and G are harmonic in Ω_ϵ. Since G vanishes on $\partial\Omega$, we get

$$(12.11) \qquad \int_{\partial\Omega} u\frac{\partial G}{\partial n}\,dS + \int_{|x-x_0|=\epsilon} u\frac{\partial G}{\partial n}\,dS - \int_{|x-x_0|=\epsilon} G\frac{du}{dn}\,dS = 0.$$

Use spherical coordinates (r,θ,φ) with pole at x_0. Then

$$G = \frac{1}{r} + H, \quad H \text{ smooth at } r = 0.$$

On $|x-x_0| = \epsilon$,

$$\frac{\partial G}{\partial n} = -\frac{\partial G}{\partial r} = \frac{1}{r^2} - \frac{\partial H}{\partial r} = \frac{1}{\epsilon^2} + O(1).$$

So

$$\int_{|x-x_0|=\epsilon} u \frac{\partial G}{\partial n} dS = u(x_0) \cdot 4\pi + o(1)$$

$$\int_{|x-x_0|=\epsilon} G \frac{\partial u}{\partial n} dS = o(1).$$

So (12.11) yields in the limit $\epsilon = 0$,

$$\int_{\partial\Omega} u \frac{\partial G}{\partial n} dS + 4\pi u(x_0) = 0,$$

i.e., (12.5) holds.

Exercise 12.1: Let Ω be the ball $|x-\bar{x}| < R$.

(a) Find the Green's function for Ω with pole at $\bar{x}$.

(b) Show that (12.5) with $x_0 = \bar{x}$ yields the mean-value property for harmonic functions.

Problem: Let B be the unit ball $|x| < 1$ in $\mathbb{R}^3$. For an arbitrary point x_0 in B let us find the Green's function G_{x_0} and write formula (12.5) explicitly.

In the solution of this problem, two ideas are used: The potential of a unit charge at x_0 satisfies (12.2) and (12.3), but fails to satisfy (12.1). Let us hunt for a negative point charge e placed at a point x_1 outside B such that the combined potential of the two charges

$$\frac{1}{|x-x_0|} + \frac{e}{|x-x_1|}$$

satisfies (12.1), i.e., vanishes on $|x| = 1$.

The second idea is geometrical: let α be a point in B. The <u>reflection</u> of α in the sphere ∂B is the point α^* such that

(i) The ray from 0 to α passes through α^*.

(ii) $|\alpha|\cdot|\alpha^*| = 1$.

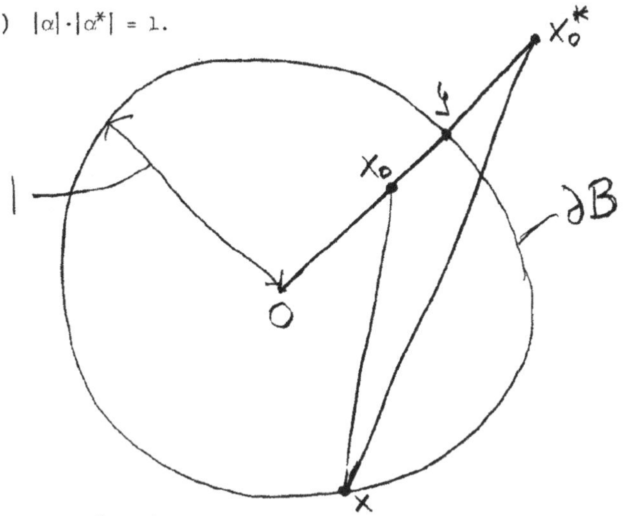

<u>Assertion</u>: The ratio $\dfrac{|x-x_0|}{|x-x_0^*|}$ is constant as x varies over $|x| = 1$.

For

$$|x-x_0|^2 = |x|^2 + |x_0|^2 - 2(x,x_0)$$
$$= 1 + |x_0|^2 - 2(x,x_0) \quad \text{if} \quad |x| = 1.$$

Similarly,

$$|x-x_0^*|^2 = 1 + |x_0^*|^2 - 2(x,x_0^*) \quad \text{if} \quad |x| = 1.$$

By (i) and (ii), $x_o^* = \dfrac{x_o}{|x_o|^2}$, so

$$|x-x_o^*|^2 = 1 + \frac{1}{|x_o|^2} - \frac{2}{|x_o|^2}(x,x_o)$$
$$= \frac{1}{|x_o|^2}\left\{|x_o|^2 + 1 - 2(x,x_o)\right\}.$$

Hence

$$\frac{|x-x_o|}{|x-x_o^*|} = |x_o|, \tag{12.12}$$

proving the assertion. Put now

$$G(x) = \frac{1}{|x-x_o|} + \frac{e}{|x-x_o^*|}.$$

In order that G vanish on $|x| = 1$, we need

$$\frac{1}{|x-x_o|} = -\frac{e}{|x-x_o^*|}, \text{ or } \frac{|x-x_o^*|}{|x-x_o|} = -e.$$

Putting $e = -\dfrac{1}{|x_o|}$, we thus get from (12.12), that

$$G(x) = \frac{1}{|x-x_o|} - \frac{1}{|x_o|}\frac{1}{|x-x_o^*|} \tag{12.13}$$

satisfies (12.1) also, and hence is the Green's function with pole at x_o.

Let us next compute the normal derivative of G on the boundary $|x|$ We introduce spherical coordinates (r,θ,φ) with pole at 0.

x_0 has spherical coordinates (r_0,θ_0,φ_0), so x_0^* is $(\frac{1}{r_0},\theta_0,\varphi_0)$. We fix some $x \in B$, $x = (r,\theta,\varphi)$ and denote by $\gamma = \gamma(r,\theta,\varphi)$ the angle between the segments $0x$ and $0x_0$. We view x_0 as fixed and x, and hence r,θ,φ, as variable. (12.13) and the law of cosines give

$$G(x) = \frac{1}{\sqrt{r^2 + r_0^2 - 2rr_0\cos\gamma}} - \frac{1}{r_0}\frac{1}{\sqrt{r^2 + \frac{1}{r_0^2} - 2\frac{r}{r_0}\cos\gamma}}$$

$$\frac{\partial G}{\partial r} = -(r^2 + r_0^2 - 2rr_0\cos\gamma)^{-3/2}(r - r_0\cos\gamma)$$

$$+ \frac{1}{r_0}\left(r^2 + \frac{1}{r_0^2} - 2\frac{r}{r_0}\cos\gamma\right)^{-3/2}\left(r - \frac{1}{r_0}\cos\gamma\right)$$

Now put $r = 1$ and note $\frac{\partial G}{\partial r} = \frac{\partial G}{\partial n}$.

$$\frac{\partial G}{\partial n}(1,\theta,\varphi) = -(1 + r_0^2 - 2r_0\cos\gamma)^{-3/2}(1 - r_0\cos\gamma)$$

$$+ \frac{1}{r_0}\left(\frac{r_0^2 + 1 - 2r_0\cos\gamma}{r_0^2}\right)^{-3/2}\left(1 - \frac{1}{r_0}\cos\gamma\right)$$

$$= -(1 + r_0^2 - 2r_0\cos\gamma)^{-3/2}(1 - r_0\cos\gamma)$$

$$+ (1 + r_0^2 - 2r_0\cos\gamma)^{-3/2}(r_0^2 - r_0\cos\gamma)$$

$$= (1 + r_0^2 - 2r_0\cos\gamma)^{-3/2}(r_0^2 - 1).$$

For x on $|x| = 1$, $|x-x_0| = (1 + r_0^2 - 2r_0\cos\gamma)^{1/2}$, so for such an x

$\frac{\partial G}{\partial n}(x) = \frac{|x_0|^2 - 1}{|x-x_0|^3}$. (12.5) gives

Theorem 12.2: u harmonic in $|x| < 1$, continuous in $|x| \leq 1$. x_o is a point in $|x| < 1$. Then

(12.14)
$$u(x_o) = \frac{1}{4\pi} \int_{|x|=1} u(x) \cdot \frac{1-|x_o|^2}{|x-x_o|^3} dS.$$

Note: Formula (12.14) is due to Poisson, dating from 1820.

Formula (12.14) can be turned around. Begin with a function f, defined and continuous on $|x| = 1$. Put

$$F(x_o) = \frac{1}{4\pi} \int_{|x|=1} f(x) \frac{1-|x_o|^2}{|x-x_o|^3} dS.$$

Exercise 12.2: Show

(a) For fixed x on $|x| = 1$, $\frac{1-|x_o|^2}{|x-x_o|^3}$ is a harmonic function of x_o for $|x_o| < 1$.

(b) F is harmonic in $|x_o| < 1$.

(c) F is continuous in $|x_o| \leq 1$ and coincides with f on $|x| = 1$.

Note: This Exercise verifies Proposition 11.1 in the special case when the region W is the unit ball.

Definition 12.2: Let Ω be a bounded domain. A function G defined on $\Omega \times \Omega$ is called the Green's function of Ω if $G(x,y)$ is the Green's function of Ω, as function of x, with pole at y, i.e., if

$$G(x,y) = G_y(x), \quad x,y \in \Omega.$$

Evidently, G is singular on the diagonal $x = y$.

Exercise 12.3: $G(x,y) > 0$ for all x,y.

Exercise 12.4: The Green's function of the unit ball is, in virtue of (12.13)

$$G(x,y) = \frac{1}{|x-y|} - \frac{1}{|y|} \cdot \frac{1}{|x-y^*|} .$$

Show by direct calculation that G is symmetric, i.e., $G(x,y) = G(y,x)$.

Exercise 12.5: Generalize the result of the last exercise to arbitrary smoothly bounded domains which possess a Green's function smooth on $\bar{\Omega} \times \bar{\Omega}$ (except on the diagonal), i.e., show the Green's function is symmetric.

Exercise 12.6: Extend the notion of Green's function to bounded domains in $\mathbb{R}^n$, $n > 3$, proceeding by analogy. Show the results of the present section have natural generalizations to $n > 3$.

Exercise 12.7: By making a suitable change of variable in (12.13), show that the Green's function of the ball $\{x| \ |x| < \rho\}$ equals

$$\frac{1}{|x-y|} - \frac{\rho}{|y|} \frac{1}{\left|x - \frac{\rho^2 y}{|y|^2}\right|} .$$

13. The Kelvin Transform

The transformation τ of $\mathbb{R}^2\setminus\{0\}$ into itself which is reflection in the un: circle:

$$\tau(z) = \frac{1}{\bar{z}}$$

is well-known to have the following properties:

(i) $\tau\circ\tau$ = identity.

(ii) If u is harmonic in a region Ω, then $u(\tau)$ is harmonic in $\tau^{-1}(\Omega) = \tau(\Omega)$.

(iii) τ preserves the magnitude of angles, reversing their sign.

Lord Kelvin, in 1847, generalized these considerations to $\mathbb{R}^3$, and his results go over to all $\mathbb{R}^n$, $n > 2$. It turns out that condition (ii) changes somewhat.

Fix n. Let T denote reflection in the unit sphere in $\mathbb{R}^n$:

$$Tx = \frac{x}{|x|^2}.$$

Consider a region Ω in $\mathbb{R}^n$, $0 \notin \Omega$, and $T\Omega$ its image-region under T. Fix a smooth function U in Ω and let F be the corresponding function on $T\Omega$, i.e.,

$$U(x) = F(Tx) \quad \text{for} \quad x \in \Omega.$$

Theorem 13.1:

(13.1) $$\Delta U(x) = |x'|^{n+2} \sum_{i=1}^{n} \frac{\partial^2}{\partial x_i'^2}\left\{\frac{1}{|x'|^{n-2}} \cdot F(x')\right\},$$

for $x \in \Omega$, the right side being evaluated for $x' = Tx$.

Corollary 1: U is harmonic in Ω if and only if $\frac{1}{|y|^{n-2}} F(y)$ is harmonic for $y \in T\Omega$.

Corollary 2: If U is defined in a region Ω of $\mathbb{R}^3$, U is harmonic in Ω if and only if $\frac{1}{|y|} U(Ty)$ is harmonic in $T\Omega$.

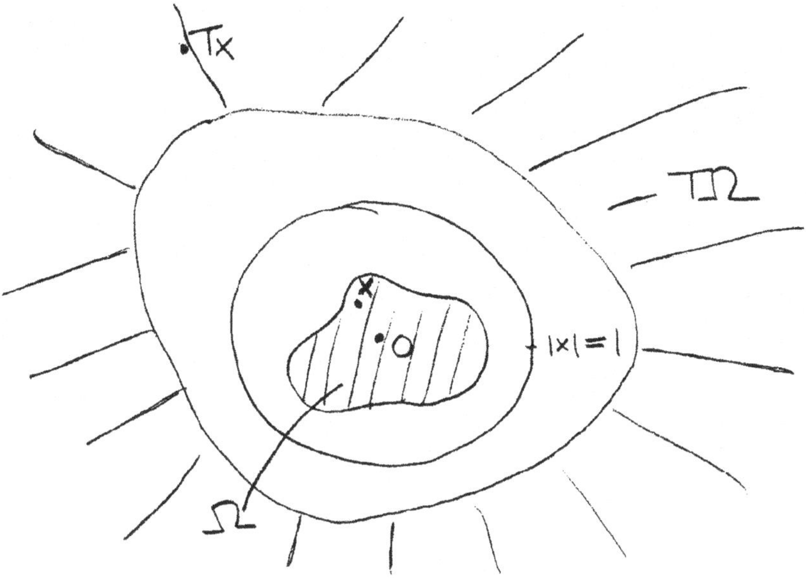

Introduce the functions x_i':

$$x_i'(x) = \frac{x_i}{|x|^2} .$$

Then $Tx = (x_1', \ldots, x_n')$. Put $r = |x|$, $r' = |x'|$.

Exercise 13.1:

(13.2)
$$\sum_{\alpha=1}^{n} \frac{\partial x_i'}{\partial x_\alpha} \cdot \frac{\partial x_j'}{\partial x_\alpha} = \begin{cases} r^{-4} & \text{if } i = j \\ 0 & \text{if } i \neq j \end{cases} .$$

Exercise 13.2:

(13.3)
$$\sum_{\alpha=1}^{n} \frac{\partial^2 x_i'}{\partial x_\alpha^2} = -2(n-2) \frac{x_i}{r^4} .$$

Exercise 13.3: If a,b are smooth functions on a region in $\mathbb{R}^n$,

(13.4)
$$\Delta(a\cdot b) = \Delta a + 2 \sum_{i=1}^{n} \frac{\partial a}{\partial x_i} \cdot \frac{\partial b}{\partial x_i} + \Delta b.$$

Proof of Theorem 13.1:

$$U(x) = F(Tx) = F(x_1',\dots,x_n').$$

Hence

(13.5)
$$\Delta U(x) = \sum_{i,j} \frac{\partial^2 F}{\partial x_i' \partial x_j'} \left(\sum_{\alpha=1}^{n} \frac{\partial x_i'}{\partial x_\alpha} \cdot \frac{\partial x_j'}{\partial x_\alpha} \right) + \sum_i \frac{\partial F}{\partial x_i'} \left(\sum_{\alpha=1}^{n} \frac{\partial^2 x_i'}{\partial x_\alpha^2} \right).$$

By (13.2) and (13.3) this gives

$$\Delta U(x) = r^{-4} \cdot \sum_i \frac{\partial^2 F}{\partial x_i'^2} - 2(n-2) \sum_i \frac{\partial F}{\partial x_i'} \cdot \frac{x_i}{r^4} .$$

We note that

$$r' = \frac{1}{r}, \quad \frac{x_i}{r^4} = x_i' \cdot r'^2,$$

(13.6) $$\Delta U(x) = r'^4 \sum_i \frac{\partial^2 F}{\partial x_i'^2} - (n-2) \sum_i \frac{\partial F}{\partial x_i'} \cdot 2x_i' r'^2$$

Denoting $\sum_i \frac{\partial^2}{\partial x_i'^2}$ by Δ' and using (13.4), we have

$$\Delta'\left(\frac{1}{r'^{n-2}} F\right) = \frac{1}{r'^{n-2}} \cdot \Delta' F + 2 \sum_i \frac{\partial F}{\partial x_i'} \cdot \frac{\partial}{\partial x_i'}\left(\frac{1}{r'^{n-2}}\right) ,$$

where we also used that $\Delta'\left(\frac{1}{r'^{n-2}}\right) = 0$. Now

$$\frac{\partial}{\partial x_i'} (r'^{2-n}) = (2-n) r'^{1-n} \frac{x_i'}{r'} = (2-n) \frac{x_i'}{r'^n} ,$$

so

$$\Delta'\left(\frac{1}{r'^{n-2}} F\right) = \frac{1}{r'^{n-2}} \Delta' F + 2 \sum_i \frac{\partial F}{\partial x_i'} (2-n) \frac{x_i'}{r'^n} .$$

Multiplying the last equation by r'^{n+2} gives

$$r'^{n+2} \Delta'\left(\frac{1}{r'^{n-2}} F\right) = r'^4 \Delta' F + 2 \sum_i \frac{\partial F}{\partial x_i'} (2-n) x_i' r'^2 .$$

Comparing this with (13.6), we have

$$\Delta U(x) = r'^{n+2} \cdot \Delta'\left(\frac{1}{r'^{n-2}} F\right) ,$$

i.e., (13.1), and we are done.

<u>Theorem 13.2</u>: <u>Let</u> Ω <u>be a bounded domain in</u> $\mathbb{R}^n$ <u>whose boundary is a finite union of smooth compact</u> (n-1)-<u>manifolds</u>. <u>Fix</u> $x_0 \in \Omega$. <u>Then</u> Ω <u>has a Green's function with pole at</u> x_0.

Proof: Since the maps:

$$u(x) \to u(x-a) \quad \text{and} \quad u(x) \to u(t \cdot x),$$

where $a \in \mathbb{R}^3$, $t \in R$, preserve the class of harmonic functions, it is no loss of generality if, in our proof, we assume that $\Omega \subset$ unit ball in $\mathbb{R}^3$ and that $x_0 = 0$.

Let $\Omega_0 = \Omega \setminus \{0\}$ and $\Omega'_0 = T\Omega$. To each function V harmonic in Ω'_0, we can associate a function U harmonic in Ω_0, as follows: put

$$(13.7) \qquad U(x) = |Tx| \cdot V(Tx), \ x \in \Omega_0.$$

Then

$$\frac{U(Ty)}{|y|} = V(y), \quad y \in \Omega'_0$$

Since $\Delta V = 0$, Corollary 2 above yields that U is harmonic in Ω_0.

If, in addition, $V = 0$ on $\partial\Omega'_0$, then $U = 0$ on $\partial\Omega$, since $T(\partial\Omega) = \partial\Omega'_0$.

Ω'_0 is the complement of a certain compact set E. The boundary of $E = T(\partial\Omega)$. Since $\partial\Omega$ is smooth, and T is a smooth map, E satisfies the cone condition (recall Definition (10.2)) at each point.

If μ is the equilibrium distribution for E, then $U^\mu = \lambda$ everywhere on E, where λ is a constant $\neq 0$.

Put $V = U^\mu - \lambda$. Then V is harmonic on Ω'_0 and $V = 0$ on $\partial\Omega'_0$. Define U by (13.7). Then U is harmonic in Ω_0 and $U = 0$ on $\partial\Omega$.

Setting $x' = Tx$, we have for $x \in \Omega_0$:

$$U(x) = |x'|\{U^\mu(x') - \lambda\},$$

so

$$U(x) + \lambda|x'| = |x'|U^\mu(x'),$$

or

$$U(x) + \lambda \frac{1}{|x|} = |x'|U^{*}(x').$$

It is clear that the right side in the last equation is bounded for large $|x'|$, whence $U(x) + \lambda \frac{1}{|x|}$ is bounded in a deleted neighborhood of the origin. By Section 11, $U(x)+\lambda\frac{1}{|x|}$ has a removable singularity at the origin. Put

$$\mathfrak{G}(x) = -\frac{1}{\lambda} U(x).$$

Then $\mathfrak{G}$ is harmonic in $\Omega\setminus\{0\}$, $\mathfrak{G}$ vanishes on $\partial\Omega$, and

$$\mathfrak{G}(x) - \frac{1}{|x|} = -\frac{1}{\lambda}\{U(x) + \frac{\lambda}{|x|}\}$$

is harmonic at 0. Hence $\mathfrak{G}$ is the Green's function for Ω with pole at 0, and we are done.

Let Ω be a bounded region in $\mathbb{R}^n$. The <u>Dirichlet Problem</u> for Ω is the problem of finding, for each continuous function f on $\partial\Omega$, a corresponding harmonic function F on Ω such that F assumes the boundary values f continuously on $\partial\Omega$.

If the Dirichlet problem for Ω is solvable, then the Green's function G_{x_0} for Ω exists for each $x_0 \in \Omega$. For, fix $x_0 \in \Omega$ and put $f(x) = -\frac{1}{|x-x_0|}$. Then f is continuous on $\partial\Omega$, so $\exists\, F$ harmonic in Ω with

$$F(x) = -\frac{1}{|x-x_0|}, \qquad x \in \partial\Omega.$$

Put $G_{x_0}(x) = F(x) + \frac{1}{|x-x_0|}$. Then G_{x_0} evidently satisfies (12.1), (12.2), (12.3), as desired. A partial converse is given by the following:

<u>Theorem 13.3: Let Ω be a bounded domain in R^n such that $\partial\Omega$ has n-dimensional measure zero. If Ω has a Green's function G_{x_0} for every x_0 in Ω, then the Dirichlet problem for Ω is solvable.</u>

Proof: Let B the subspace of $C(\partial\Omega)$ consisting of those functions which are boundary values, i.e. $f \in B$ if and only if $\exists F$ harmonic in Ω and continuous in $\bar{\Omega}$ with $F = f$ on $\partial\Omega$.

Because of the maximum principle, B is a closed subspace of $C(\partial\Omega)$, as the reader may easily verify.

The assertion of the theorem amounts to showing that $B = C(\partial\Omega)$. If $B \neq C(\partial\Omega)$, by F. Riesz $\exists$ signed measure α on $\partial\Omega$ such that

(13.8) $$\int f d\alpha = 0, \quad \text{all } f \in B$$

and with $\alpha \neq 0$.

Fix $x_0 \in \Omega$. By hypothesis, G_{x_0} exists and $G_{x_0}(x) = \frac{1}{|x-x_0|} + H(x)$, where H is harmonic in Ω and continuous on $\bar{\Omega}$. Since $G_{x_0} = 0$ on $\partial\Omega$, $H(x) = -\frac{1}{|x-x_0|}$ on $\partial\Omega$, and so $\frac{1}{|x-x_0|} \in B$. For $x_0 \in \mathbb{R}^3\setminus\bar{\Omega}$, $\frac{1}{|x-x_0|} \in B$ since the function is harmonic in a neighborhood of $\bar{\Omega}$. By (13.8), then

$$\int \frac{d\alpha(x)}{|x-x_0|} = 0, \quad x_0 \in \mathbb{R}^3\setminus\partial\Omega,$$

and so a.e. in $\mathbb{R}^3$. But $\alpha = \alpha_1 - \alpha_2$, where α_1, α_2 are measures, and so

$$U^{\alpha_1} = U^{\alpha_2} \quad \text{a.e.}$$

It follows, as on earlier occasions, that $\alpha_1 = \alpha_2$ or $\alpha = 0$, which is a contradiction. Thus $B = C(\partial\Omega)$, and we are done.

Corollary: Let Ω be a smoothly bounded domain in $\mathbb{R}^n$, as in Theorem 13.2. Then the Dirichlet problem is solvable for Ω.

Proof: By Theorem 13.2 Ω has a Green's function for each $x_0 \in \Omega$. By Theorem 13.3 this gives the assertion, since $\partial\Omega$ surely has n-dimensional measure 0, being a union of finitely many smooth manisfolds.

Note: Proposition 11.1 is now proved.

14. Perron's Method

In the last section we saw a way of solving the Dirichlet problem for smoothly bounded domains. We shall now consider the same problem for an arbitrary bounded domain, making use of a method given by O. Perron in 1923.

A simple-minded approach to the problem might be this: let Ω be a bounded region in $\mathbb{R}^n$ and let $f \in C(\partial\Omega)$. We seek a function Φ_f continuous in $\bar{\Omega}$, harmonic in Ω, and with $\Phi_f = f$ on $\partial\Omega$.

Surely $\exists$ functions V harmonic on Ω with $V \leq f$ on $\partial\Omega$. Put

$$\mathfrak{S}_f = \{V \text{ harmonic in } \Omega \mid V \leq f \text{ on } \partial\Omega\}.$$

If Φ_f exists, then $\Phi_f \in \mathfrak{S}_f$, and whenever $V \in C_f$, $V \leq \Phi_f$ on Ω. Hence for all $x \in \Omega$,

$$\Phi_f(x) = \sup_{V \in \mathfrak{S}_f} V(x). \tag{14.1}$$

This suggests that, to construct Φ_f we define Φ_f by (14.1). If we do this, our first problem is to show that Φ_f, as defined in (14.1), is harmonic in Ω.

Given a family $\mathfrak{S}$ of functions harmonic in Ω and uniformly bounded, and put

$$F(x) = \sup_{A \in \mathfrak{S}} A(x), \quad \text{all } x \in \Omega.$$

Is F harmonic?

Consider the case when $n = 1$ and $\mathfrak{S}$ consists of the two functions $A_1(x) = x$, $A_2(x) = -x$. A_1 and A_2 are harmonic on $-1 < x < 1$, yet $\sup_{A \in \mathfrak{S}} A(x) = |x|$, which fails to be harmonic. However, $|x|$ is convex!

Let ω be a finite-interval on $\mathbb{R}^1$ and $\mathfrak{K}$ the class of convex functions on ω. Then

(i) If $a,b \in \mathfrak{K}$, then $\max(a,b) \in \mathfrak{K}$.

(ii) Let I be a closed interval $\subset \omega$ and h harmonic (thus linear) on I. If $u \in \mathfrak{K}$ and $u \leq h$ on ∂I, then $u \leq h$ on I.

(iii) Let I be as above and $f \in \mathfrak{K}$. The function $\tilde{f}$ which agrees with f on $\omega \setminus \dot{I}$ and is harmonic on I again $\in \mathfrak{K}$.

We generalize this situation now, with ω replaced by a bounded domain $\Omega \subset \mathbb{R}^n$ and $\mathfrak{K}$ replaced by a class $\mathfrak{S}$ of continuous functions defined on Ω.

Notation: If B is a closed ball $\subset \Omega$ and $\varphi \in \mathfrak{S}$, $(\varphi)_B$ denotes the function which coincides with φ on $\Omega \setminus \dot{B}$ and is harmonic in $\dot{B}$.

We impose the following conditions on $\mathfrak{S}$, generalizing (i) to (iii) above.

(14.2) If $a,b \in \mathfrak{S}$, then $\max(a,b) \in \mathfrak{S}$.

(14.3) Let B be a closed ball $\subset \Omega$ and let h be harmonic on $\dot{B}$, continuous on B. If $u \in \mathfrak{S}$ and $u \leq h$ on ∂B, then $u \leq h$ on B.

(14.4) Let B be as above and let $\varphi \in \mathfrak{S}$. Then $(\varphi)_B \in \mathfrak{S}$.

Note: For $\varphi \in \mathfrak{S}$, $(\varphi)_B \geq \varphi$ everywhere on Ω. For $(\varphi)_B = \varphi$ outside $\dot{B}$; also $\varphi \in \mathfrak{S}$ and $\varphi \leq (\varphi)_B$ on ∂B and $(\varphi)_B$ is harmonic on $\dot{B}$. So by (14.3),

$$\varphi \leq (\varphi)_B \text{ on } B.$$

In addition, we assume

(14.5) $\exists M$ such that $u \leq M$ on Ω for all $u \in \mathfrak{S}$.

Theorem 14.1: *Let* $\mathfrak{S}$ *satisfy* (14.2) *through* (14.5). *Put, for* $x \in \Omega$,

(14.6) $$U(x) = \sup_{u\in\mathfrak{S}} u(x).$$

Then U _is harmonic in_ Ω.

Proof: Fix a closed ball $B \subset \Omega$ and fix $x_o \in \dot{B}$.

Choose a sequence $\{\varphi_n\}$ in $\mathfrak{S}$ such that $\varphi_n(x_o) \uparrow U(x_o)$. Put

$$\psi_1 = (\varphi_1)_B.$$

Then $\varphi_1 \le \psi_1$, $\psi_1 \in \mathfrak{S}$ and ψ_1 is harmonic in $\dot{B}$. Put

$$\rho_2 = \max(\psi_1, \varphi_2)$$

and

$$\psi_2 = (\rho_2)_B.$$

Then $\rho_2 \in \mathfrak{S}$, so $\psi_2 \in \mathfrak{S}$.

$$\psi_1 \le \rho_2 \le \psi_2$$

and

$$\varphi_2 \le \rho_2 \le \psi_2.$$

Also ψ_2 is harmonic in $\dot{B}$.

Suppose we have found $\psi_1, \ldots, \psi_n \in \mathfrak{S}$ such that each ψ_j is harmonic in $\dot{B}$, and for all j

$$\psi_j \le \psi_{j+1} \text{ and } \varphi_j \le \psi_j.$$

Put $\rho_{n+1} = \max(\psi_n, \varphi_{n+1})$, and $\psi_{n+1} = (\rho_{n+1})_B$. Then $\psi_{n+1} \in \mathfrak{S}$, and

$$\psi_n \le \rho_{n+1} \le \psi_{n+1}$$

and

$$\varphi_{n+1} \le \rho_{n+1} \le \psi_{n+1}.$$

Hence we have extended the ψ_j to a set $\psi_1,\ldots,\psi_{n+1}$ with the same properties. By induction there hence exists a sequence $\{\psi_n | 1 \leq n < \infty\}$ in $\mathfrak{S}$ with

(14.7) $$\psi_n \leq \psi_{n+1}, \text{ all } n.$$

(14.8) $$\varphi_n \leq \psi_n, \text{ all } n.$$

(14.9) $$\psi_n \text{ harmonic in } \dot{B}, \text{ all } n.$$

Put

$$\psi(x) = \lim_{n \to \infty} \psi_n(x).$$

ψ exists for all x in Ω by (14.7) and (14.5). ψ is harmonic in $\dot{B}$, because by Harnack's theorem a bounded monotone sequence of harmonic functions has a harmonic limit. Also $\psi(x_0) = U(x_0)$, by (14.8).

<u>Assertion</u>: $\psi = U$ in $\dot{B}$.

Fix $x_1 \in \dot{B}$. Choose $\{\alpha_n\} \in \mathfrak{S}$ such that $\alpha_n(x_1) \uparrow U(x_1)$. Put

$$\beta_1 = \max(\alpha_1,\psi_1), \quad \gamma_1 = (\beta_1)_B,$$

where ψ_1 is as above. $\gamma_1 \in \mathfrak{S}$ and $\gamma_1 \geq \alpha_1$, $\gamma_1 \geq \psi_1$, γ_1 is harmonic in $\dot{B}$.

Having chosen $\gamma_1,\ldots,\gamma_n \in \mathfrak{S}$ such that for each j

(i) γ_j is harmonic in $\dot{B}$,

(ii) $\gamma_j \geq \alpha_j$,

(iii) $\gamma_j \geq \psi_j$,

(iv) $\gamma_j \geq \gamma_{j-1}$,

we put $\beta_{n+1} = \max(\gamma_n,\psi_{n+1},\alpha_{n+1})$, and $\gamma_{n+1} = (\beta_{n+1})_B$. Clearly γ_{n+1} satisfies (i) through (iv) with $j = n+1$. By induction there thus exists an infinite sequence

$\{\gamma_j\}$ in $\mathfrak{S}$ satisfying (i) through (iv).

Put

$$\gamma(x) = \lim_{n\to\infty} \gamma_n(x), \quad x \in \Omega.$$

Then, arguing as above, we see that γ is harmonic in $\dot{B}$. Also by (iii), $\gamma \geq \psi$ and $\gamma(x_0) \leq U(x_0) = \psi(x_0)$. Since ψ is also harmonic in $\dot{B}$, $\gamma - \psi$ is a non-negative harmonic function in $\dot{B}$ which vanishes at x_0, and so vanishes identically in $\dot{B}$. Also

$$\gamma(x_1) \geq \lim_{n\to\infty} \alpha_n(x_1) = U(x_1).$$

Hence $\gamma(x_1) = U(x_1)$, for $\gamma \leq U$ since each $\gamma_n \in \mathfrak{S}$. Hence $\psi(x_1) = U(x_1)$.

But x_1 was arbitrary in $\dot{B}$, so our assertion is proved. Hence U is harmonic in $\dot{B}$. Since B was arbitrary, U is harmonic in Ω. q.e.d.

We shall obtain a class of functions which satisfy (14.2) through (14.5) by generalizing in a suitable way the notion of a convex function one real variable.

Let ω be an interval on $\mathbb{R}$ and f a convex function defined on ω. For each $x_0 \in \omega$ and each r such that $(x_0-r, x_0+r) \subset \omega$, we have

$$f(x_0) \leq \frac{1}{2}\{f(x_0-r) + f(x_0+r)\}. \tag{14.10}$$

The right side in (14.10) is the mean value of f on the boundary of the interval (x_0-r, x_0+r). Let now n be arbitrary, $x_0 \in \mathbb{R}^n$, $r > 0$.

<u>Definition 14.1</u>: For f a function defined on $|x-x_0| = r$,

$$M(f;x_0,r) = \frac{1}{A} \int_{|x-x_0|=r} f\,dS,$$

where A is the area of $\{x \mid |x-x_0| = r\}$.

Let Ω be a region in $\mathbb{R}^n$ and u a function defined on Ω. Our generalization of (14.10) is this:

$$u(x_o) \leq M(u;x_o,r). \tag{14.11}$$

We make the following definition.

Definition 14.2: A function u is subharmonic in Ω if (14.11) holds for each $x_o \in \Omega$ and every $r < r(x_o)$, where $r(x_o) > 0$, and if u is upper semi-continuous, i.e.

$$\overline{\lim_{x \to x_o}}\ u(x) \leq u(x_o), \quad \text{all } x_o \in \Omega.$$

We allow u to take the value $-\infty$, but not $+\infty$. Also u may not be identically $-\infty$ on any ball.

Convex functions on $\mathbb{R}^1$ are "sublinear", i.e. if f is convex and I is an interval, and if h is a linear function with $f \leq h$ on ∂I, then $f \leq h$ on I. Subharmonic functions enjoy the corresponding property, with "linear" replaced by "harmonic".

Theorem 14.2: Let Ω be a bounded domain in $\mathbb{R}^n$ and u subharmonic in Ω. Let ω be an open set with $\overline{\omega} \subset \Omega$ and let h be continuous in $\overline{\omega}$, harmonic in ω. If

$$u \leq h \text{ on } \partial\omega, \text{ then } u \leq h \text{ on } \omega.$$

Lemma 14.3: Let ω be a bounded domain in $\mathbb{R}^n$ and u subharmonic in ω. If $\overline{\lim_{x \to x_o}}\ u(x) \leq 0$ for all $x_o \in \partial\omega$, then $u \leq 0$ in ω.

The Lemma implies the Theorem, since $u - h$ is subharmonic when u is subharmonic and h is harmonic. The Lemma is proved by the standard argument used in proving the maximum principle for harmonic functions. The argument works just as well for subharmonic functions, and the reader should supply the details.

Exercise 14.1: Ω a domain in $\mathbb{R}^n$. Let a,b be subharmonic in Ω. Then $\max(a,b)$ is subharmonic in Ω.

Lemma 14.4: *Ω is a domain in $\mathbb{R}^n$ and u is continuous and subharmonic in Ω. B is a closed ball $\subset \Omega$. $(u)_B$ is defined as the continuous function on Ω coinciding with u outside $\dot{B}$ and harmonic in $\dot{B}$. Then $(u)_B$ is subharmonic in Ω.*

Proof: Since $(u)_B$ is subharmonic on $\Omega\setminus B$ and on $\dot{B}$, we only have to verify (14.11) for points $x_0 \in \partial B$.

Fix such a point x_0.

$$(u)_B(x_0) = u(x_0) \le \frac{1}{4\pi R^2} \int_{|x-x_0|=R} u\,dS, \tag{14.12}$$

for small R. Since $(u)_B$ is harmonic in $\dot{B}$ and $u = (u)_B$ on ∂B, Theorem 14.2 gives $u \le (u)_B$ on $\dot{B}$. Hence $u \le (u)_B$ everywhere, and so (14.12) yields

$$(u)_B(x_0) \le \frac{1}{4\pi R^2} \int_{|x-x_0|=R} (u)_B\,dS,$$

for small R. Thus $(u)_B$ is subharmonic in Ω. q.e.d.

Let now Ω be a bounded domain in $\mathbb{R}^n$ and f a bounded function defined on $\partial\Omega$. Define

$$\mathfrak{S}_f = \{u \mid u \text{ is continuous and subharmonic on } \Omega \text{ and for all } x_0 \in \partial\Omega,\ \overline{\lim_{x \to x_0}}\, u(x) \le f(x_0)\}.$$

Assertion: $\mathfrak{S}_f$ satisfies conditions (14.2) through (14.5).

Proof: Exercise 14.1 yields (14.2). Theorem 14.2 yields (14.3). Lemma 14.4 yields (14.4). Since f is bounded on $\partial\Omega$, $\exists M$ such that

$$\overline{\lim_{x \to x_0}}\, u(x) \leq M \quad \text{for all} \quad x_0 \in \partial\Omega,\ u \in \mathfrak{S}_f.$$

Lemma 14.3 yields that $u \leq M$ in Ω. Hence (14.5) holds.

Theorem 14.5: *Let $\Omega, f, \mathfrak{S}_f$ be as above. Define*

$$U_f(x) = \sup_{u \in \mathfrak{S}_f} u(x), \quad x \in \Omega.$$

Then U_f is harmonic in Ω.

Proof: The Assertion just proved shows that Theorem 14.1 applies to $\mathfrak{S}_f$.

Note: The mean value property of harmonic functions evidently implies (14.11) whenever the ball $\{x \mid |x-x_0| \leq r\}$ is contained in the region. Thus every harmonic function in Ω is subharmonic in Ω (just as every linear function on $\mathbb{R}^1$ is convex).

Suppose the Dirichlet problem for Ω is solvable, let f be a continuous function on $\partial\Omega$ and let F be the harmonic function whose boundary function is f. Then

(14.13) $$F \in \mathfrak{S}_f$$

(14.14) $$\text{If} \quad u \in \mathfrak{S}_F, \quad u \leq F.$$

For $u - F$ is subharmonic on Ω and $\overline{\lim_{x \to x_0}}\,(u-F)(x_0) \leq 0$ if $x_0 \in \partial\Omega$. By Lemma 14.3, $u - F \leq 0$ on Ω, i.e. (14.14) holds.

Hence $U_f = F$. Thus, to show that Dirichlet's problem is solvable, we must show that U_f takes on f as its boundary function on $\partial\Omega$. Whether or not this is true turns out to depend on delicate properties of $\partial\Omega$. We shall attack this problem in the next section.

Exercise 14.2: Let u be smooth on the ball $B: \{x \mid |x-x_o| \leq R\}$ in $\mathbb{R}^3$. Show

$$(14.15) \qquad u(x_o) = \frac{1}{4\pi R^2} \int_{\partial B} u\,dS - \frac{1}{4\pi} \int_B \Delta u \left\{ \frac{1}{|x-x_o|} - \frac{1}{R} \right\} dx.$$

Deduce that a smooth function u is subharmonic in a region if and only if $\Delta u \geq 0$ everywhere.

Exercise 14.3: The class of all functions subharmonic in a region is a cone, i.e. u_1, u_2 subharmonic imply $u_1 + u_2$ subharmonic, and u subharmonic, t a positive constant implies $t \cdot u$ subharmonic.

15. Barriers

Let Ω be a bounded domain in $\mathbb{R}^n$ and f a bounded function on $\partial\Omega$. Put

$$U_f(x) = \sup u(x)$$

taken over all $u \in \mathfrak{S}_f$, where $\mathfrak{S}_f$ is the class of all subharmonic and continuous functions u on Ω which satisfy $\overline{\lim}_{x \to x_0} u(x) \leq f(x_0)$ for all $x_0 \in \partial\Omega$.

In the last section we saw that U_f is harmonic in Ω.

Definition 15.1: A point x_0 in $\partial\Omega$ is called regular provided that whenever f is a bounded function on $\partial\Omega$ which is continuous at x_0, then

$$\lim_{x \to x_0} U_f(x) = f(x_0).$$

Theorem 15.1: A necessary condition in order that x_0 be regular is that $\exists$ a harmonic function V on Ω such that

(15.1) $$\text{If } \delta > 0, \quad \inf_{|x-x_0|>\delta} V(x) > 0,$$

and

(15.2) $$\lim_{x \to x_0} V(x) = 0.$$

Proof: If every point on $\partial\Omega$ were regular, we could obtain V by choosing a continuous function f on $\partial\Omega$ with $f(x_0) = 0$ and $f > 0$ elsewhere on $\partial\Omega$, and putting $V = U_f$.

Knowing only that x_0 is regular, we proceed as follows: let $f(x) = |x-x_0|$, restricted to $\partial\Omega$. Let $\bar{u}(x) = |x-x_o|$ for $x \in \Omega$. Verify that $\bar{u}$ is subharmonic and hence $\in \mathfrak{S}_f$.

Put $V = U_f$. Then V is harmonic in Ω. Since $\bar{u} \in \mathfrak{S}_f$, $V \geq \bar{u}$ in Ω, and so (15.1) holds. Since x_o is regular, $\lim\limits_{x \to x_o} V(x) = f(x_o) = 0$, i.e., (15.2). We are done.

It turns out that the converse of the last theorem is true, i.e., the existence of a harmonic function V satisfying (15.1) and (15.2) assures that x_o is regular. In fact, a slightly weaker hypothesis suffices.

Definition 15.2: A function F is superharmonic in a region of $\mathbb{R}^n$ if $-F$ is subharmonic.

Thus harmonic functions, in particular, are superharmonic.

Definition 15.3: Let Ω be a region of $\mathbb{R}^n$ and $x_o \in \partial\Omega$. A function V is a barrier for Ω at x_o if V is continuous and superharmonic in Ω, and satisfies (15.1) and (15.2). Let Ω be a bounded region in $\mathbb{R}^n$.

Theorem 15.2: *Let $x_o \in \partial\Omega$. If $\exists$ a barrier V for Ω at x_o, then x_o is regular.*

Proof: Let f be a bounded function on $\partial\Omega$, $|f| \leq M$. Assume f is continuous at x_o. Since we can change f by a constant, we assume with no loss of generality that $f(x_o) = 0$.

Fix $\epsilon > 0$. By hypothesis, $\exists\ \rho > 0$ such that $|f(\zeta)| < \epsilon$ if $|\zeta-x_o| \leq \rho$. Let $B_\rho = \{x|\ |x-x_o| \leq \rho\}$. By definition of barrier, $\exists\ V_o > 0$ with $V(x) > V_o$ if $x \in \Omega\backslash B_\rho$. The function

$$F_\epsilon = \epsilon + \frac{M}{V_o} \cdot V$$

is superharmonic in Ω. Fix $u \in \mathfrak{S}_f$.

If $\zeta \in \partial\Omega \cap B_\rho$,

$$\varliminf_{x \to \zeta} F_\epsilon(x) \geq \epsilon \geq f(\zeta) \geq \varlimsup_{x \to \zeta} u(x). \tag{15.3}$$

If $\zeta \in \partial\Omega \backslash B_\rho$, $V > V_o$ near ζ, so

$$\varliminf_{x \to \zeta} F_\epsilon(x) \geq \epsilon + M \geq \varlimsup_{x \to \zeta} u(x). \tag{15.4}$$

Hence for all $x \in \Omega$,

$$F_\epsilon(x) \geq u(x).$$

It follows that for all $x \in \Omega$,

$$F_\epsilon(x) \geq U_f(x).$$

Letting $x \to x_o$ and using (15.2), we get

$$\epsilon \geq \varlimsup_{x \to x_o} U_f(x).$$

Since ϵ was arbitrary,

$$0 \geq \varlimsup_{x \to x_o} U_f(x).$$

On the other hand, the function

$$G_\epsilon = -\epsilon - \frac{M}{V_o} \cdot V$$

is subharmonic in Ω.

If $\zeta \in \partial\Omega \cap B_\rho$,

(15.5) $$\overline{\lim_{x \to \zeta}}\ G_\epsilon(x) \le -\epsilon \le f(\zeta).$$

If $\zeta \in \partial\Omega \setminus B_\rho$, $V > V_o$ near ζ, so

(15.6) $$\overline{\lim_{x \to \zeta}}\ G_\epsilon(x) \le -M - \epsilon \le f(\zeta).$$

Hence $G_\epsilon \in \mathfrak{S}_f$. It follows that

$$G_\epsilon(x) \le U_f(x) \quad \text{for all} \quad x \in \Omega.$$

Letting $x \to x_o$, we get

$$-\epsilon \le \underline{\lim_{x \to x_o}}\ U_f(x).$$

Since ϵ is arbitrary, we have

$$0 \le \underline{\lim_{x \to x_o}}\ U_f(x).$$

Together with the earlier bound on $\overline{\lim_{x \to x_o}}\ U_f(x)$, this yields

$$0 = \lim_{x \to x_o} U_f(x).$$

Thus we have that x_o is regular. q.e.d.

<u>Corollary 1</u>: <u>Let</u> Ω <u>be a region in</u> R^n <u>and</u> $x_o \in \partial\Omega$. <u>Assume</u> $\exists$ <u>a closed ball</u> B <u>with</u> $x_o \in \partial B$ <u>and</u> Ω <u>lying outside</u> B. <u>Then</u> x_o <u>is a regular point of</u> Ω.

Proof: Without loss of generality, ∂B meets $\partial\Omega$ only at x_0. Let $\bar{x}$ be the center and R the radius of B. Put $V(x) = \frac{1}{R} - \frac{1}{|x-\bar{x}|}$.

Then V is harmonic in Ω, continuous in $\bar{\Omega}$, $V(x_0) = 0$ and $V(x) > 0$ for $x \in \partial\Omega\setminus\{x_0\}$. Hence V is a barrier for Ω at x_0. Theorem 15.2 gives the assertion.

Note: The criterion for regularity just given is due to Poincaré, who introduced the use of barriers in 1890. The general definition of barrier was later given by Lebesgue.

In our definition of a barrier V for Ω at x_0, we demanded that $V(x) \to 0$ as $x \to x_0$ and that $V(x)$ be bounded away from 0 if x stays away from x_0. What happens if we demand instead only that $V > 0$ everywhere in Ω?

For convenience, we consider harmonic V in the next theorem.

Theorem 15.3: (Bouligand) Assume $\exists h$ harmonic in Ω with

(15.7) $$h > 0 \text{ in } \Omega.$$

(15.8) $$\lim_{x \to x_0} h(x) = 0.$$

Then $\exists H$ harmonic in Ω such that H is a barrier, i.e., satisfies (15.1) and (15.2).

Lemma 15.4: Let B be the ball $|x-x_0| < \rho$ in $\mathbb{R}^3$ and u a bounded measurable function on ∂B. Then $\exists$ U harmonic in $\dot{B}$ such that

(i) $\inf_{\partial B} u \leq U(x) \leq \sup_{\partial B} u$, $x \in \dot{B}$.

(ii) $U(x_0) = \frac{1}{4\pi\rho^2} \int_{\partial B} u dS$.

(iii) If ζ is a point of continuity of u on ∂B, then
$$\lim_{x \to \zeta} U(x) = u(\zeta).$$

Proof: By the change of variable $y = x_0 + \rho\zeta$ which carries $|\zeta| = 1$ on ∂B, we may reduce the problem to the case $x_0 = 0$, $\rho = 1$. Given u bounded and measurable on $|\zeta| = 1$, we set

$$U(x) = \frac{1}{4\pi} \int_{|\zeta|=1} u(\zeta) \frac{1-|x|^2}{|\zeta-x|^3} dS.$$

A minor modification in the proof of Exercise 12.2 then shows that U satisfies (i), (ii), (iii).

Proof of Theorem 15.3: Put $f(x) = |x-x_0|$ and $H = U_f$. As we saw in the proof of Theorem 15.1, H satisfies (15.1) and H is harmonic in Ω. It remains to show $H(x) \to 0$ as $x \to x_0$.

Fix $u \in \mathfrak{S}_f$ and $\rho > 0$. Put

$$\Omega_\rho = \{x \mid |x-x_0| < \rho\} \cap \Omega.$$

We seek an upper bound for u in Ω_ρ.

Let B be the ball $\{x \mid |x-x_0| \leq \rho\}$. Choose a compact subset e of $\partial B \cap \Omega$.

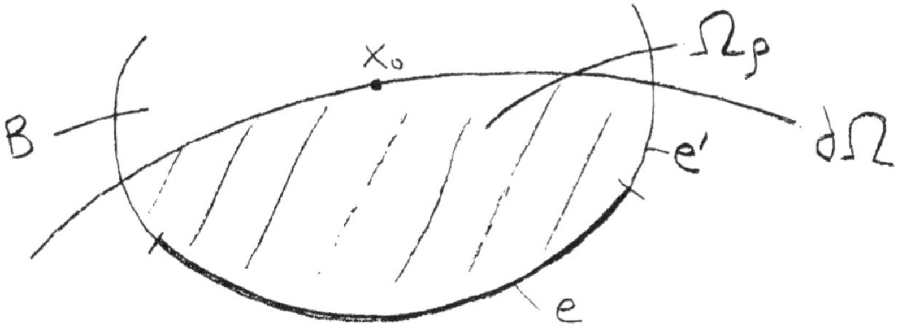

Since $h > 0$ on Ω, h has a positive minimum h_o on e. We lack a positive lower bound for h on $\partial B \backslash e$. To compensate for this, we proceed as follows:

Let $M = \max_{x \in \partial\Omega} |x - x_o|$. Let v be the function on ∂B which $= M$ on $e' = (\partial B \cap \Omega) \backslash e$, $v = 0$ on the rest of ∂B. By Lemma 15.4, $\exists\, I$ harmonic in $\dot{B}$ with $0 \leq I \leq M$ in $\dot{B}$, and

$$I(x_o) = \frac{1}{4\pi\rho^2} \int_{\partial B} v\,dS = \frac{M \text{ area}(e')}{4\pi\rho^2}. \tag{15.9}$$

$$\lim_{x \to \zeta} I(x) = M \quad \text{for all} \quad \zeta \in e'. \tag{15.10}$$

Put $\Phi(x) = \rho + \frac{M}{h_o} \cdot h + I$.

Then Φ is harmonic in Ω_ρ. We claim:

$$\underline{\lim}_{x \to \zeta} \Phi(x) \geq \overline{\lim}_{x \to \zeta} u(x) \quad \text{for all} \quad \zeta \in \partial\Omega_p.$$

For if $\zeta \in \partial\Omega \cap \partial\Omega_p$,

$$\underline{\lim}_{x \to \zeta} \Phi(x) \geq \rho \geq |\zeta - x_o| \geq \overline{\lim}_{x \to \zeta} u(x).$$

If $\zeta \in e$.

$$\underline{\lim}_{x \to \zeta} \Phi(x) \geq M \geq u(\zeta),$$

since $h \geq h_o$ on e, and $u \leq M$ everywhere, since $u \in \mathfrak{S}_f$.

If $\zeta \in e'$,

$$\underline{\lim}_{x \to \zeta} \Phi(x) \geq M \geq \overline{\lim}_{x \to \zeta} u(x),$$

because of (15.10).

Thus the claim is proved. Hence

$$u \leq \Phi \quad \text{in} \quad \Omega_\rho .$$

Since u was arbitrary in $\mathfrak{S}_f$, while Φ is independent of u,

$$H \leq \Phi \quad \text{in} \quad \Omega_\rho .$$

Now $\lim_{x \to x_0} h(x) = 0$, and $\lim_{x \to x_0} I(x) = I(x_0)$. Hence

$$\overline{\lim_{x \to x_0}} H(x) \leq \lim_{x \to x_0} \Phi(x) = \rho + I(x_0) .$$

By (15.9), we can have $I(x_0)$ as small as we please by choosing e' small enough, i.e., e big enough. We choose e so that $I(x_0) < \rho$. Then

$$\overline{\lim_{x \to x_0}} H(x\ ) \leq 2\rho .$$

Since ρ is arbitrary, this implies $\lim_{x \to x_0} H(x) = 0$. q.e.d.

16. Kellogg's Theorem

Let Ω be a bounded domain in $\mathbb{R}^n$. What can we say about the set of regular points on $\partial\Omega$?

Theorem 16.1: (Kellogg) The set of points on $\partial\Omega$ which are not regular has capacity zero.

Without loss of generality, $0 \in \Omega$ and $\bar{\Omega} \subset \{x \mid |x| < 1\}$. Recall the reflection map T: $Tx = \frac{x}{|x|^2}$. Put $\Omega' = T(\Omega\setminus\{0\})$ and $E = \mathbb{R}^3\setminus\Omega'$. E is a compact set in $\mathbb{R}^3$.

Let $\bar{\mu}$ be the equilibrium distribution for E. Then $\exists$ constant γ with $U^{\bar{\mu}} = \gamma$ on E outside of an exceptional set $\tilde{E}$. We learned in Section 7 that

(16.1) $$C(\tilde{E}) = 0.$$

(16.2) $$0 < U^{\bar{\mu}} < \gamma \quad \text{on } \mathbb{R}^3\setminus E.$$

(16.3) $$\lim_{x \to \zeta} U^{\bar{\mu}}(x) = \gamma \quad \text{if } \zeta \in E\setminus\tilde{E}.$$

Put

$$\varphi = 1 - \frac{1}{\gamma} U^{\bar{\mu}}$$

Then φ is harmonic on $\mathbb{R}^3\setminus E$,

(16.4) $$0 < \varphi \quad \text{on } \mathbb{R}^3\setminus E.$$

(16.5) $$\lim_{x \to \zeta} \varphi(x) = 0 \quad \text{if } \zeta \in E\setminus\tilde{E}.$$

We now form the harmonic function on Ω which corresponds to φ. Put

$$\Phi(x) = |Tx| \cdot \varphi(Tx), \quad x \in \Omega\setminus\{0\}.$$

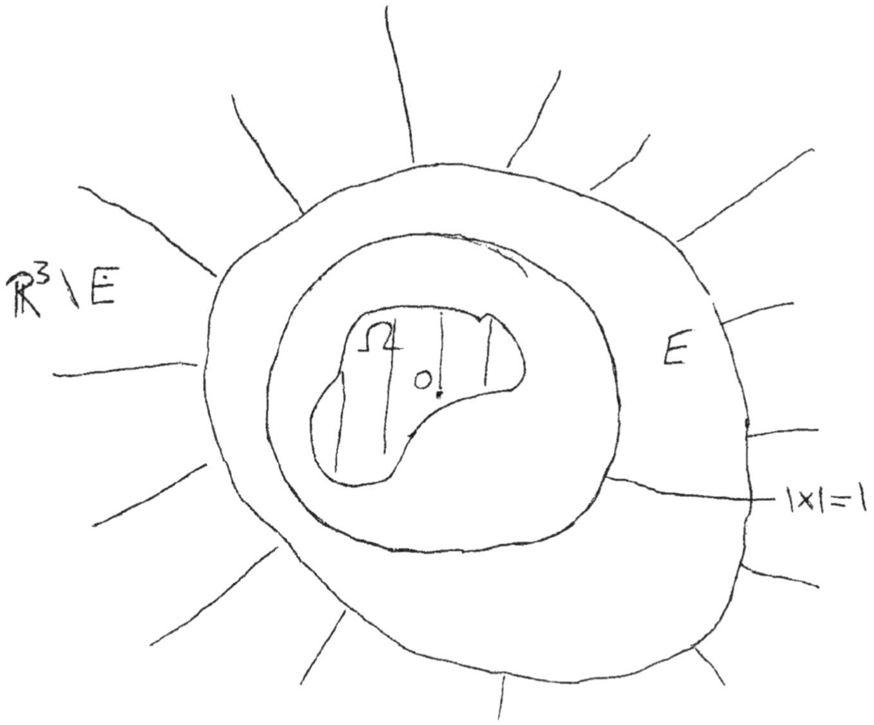

Since $\frac{1}{|x'|}\,(|x'|\varphi(x'))$ is harmonic in Ω', Φ is harmonic in $\Omega\setminus\{0\}$. By (16.4) and (16.5), $\Phi > 0$ on $\Omega\setminus\{0\}$ and $\lim\limits_{x\to x_0} \Phi(x) = 0$ if $Tx_0 \in E\setminus\tilde{E}$.

Fix $x_0 \in \partial\Omega$ such that $Tx_0 \in E\setminus\tilde{E}$.

<u>Assertion 1</u>: x_0 is a regular point of Ω.

Φ is harmonic in $\Omega\setminus\{0\}$. Theorem 15.3 yields H harmonic in $\Omega\setminus\{0\}$ such that H is a barrier for $\Omega\setminus\{0\}$ at x_0.

We now need the fact that the property of regularity of a point is a local property of the boundary.

Exercise 16.1: Let W be a bounded region in $\mathbb{R}^n$ and $\bar{x} \in \partial W$. Fix $R > 0$. Suppose $\exists$ a function S continuous and superharmonic on $\{|x-\bar{x}| < R\} \cap W$ such that

(i) If $\delta > 0$, $\inf_{|x-\bar{x}|>\delta} S(x) > 0$.

(ii) $\lim_{x \to \bar{x}} S(x) = 0$.

Then W has a barrier at $\bar{x}$.

Applying this Exercise to Ω at x_o, we see that Ω has a barrier at x_o, and so Assertion 1 holds.

Exercise 16.2: Let W be an open set in $\mathbb{R}^n$ and τ a homeomorphism of W into $\mathbb{R}^n$ such that τ and $\tau^{-1} \in C^1$. Let W_o be a compact subset of W. Then $\exists$ a positive constant M such that whenever F is a Borel set $\subset W_o$,

$$\frac{1}{M} C(F) \leq C(\tau F) \leq MC(F). \tag{16.6}$$

In particular $C(\tau F) = 0$ if and only if $C(F) = 0$.

Put now $F = T^{-1}\tilde{E}$ and apply Exercise 16.2 with $T = \tau$. Since $C(\tilde{E}) = 0$, the Exercise yields that $C(T^{-1}\tilde{E}) = 0$. For $x_o \in \partial\Omega \setminus T^{-1}\tilde{E}$, $Tx_o \in E\setminus\tilde{E}$, so Assertion 1 gives that x_o is a regular point of Ω. Thus the set of points on $\partial\Omega$ which are not regular $\subseteq T^{-1}\tilde{E}$, and so has capacity 0. Theorem 16.1 is proved.

Note: It is easy, conversely, to exhibit sets of capacity 0 which consist of irregular points. Let B be a closed ball in $\mathbb{R}^3$ and F a compact subset of $\dot{B}$ with $C(F) = 0$. Let $\Omega = \dot{B}\setminus F$. Then $F \subset \partial\Omega$.

Fix $x_o \in F$. If x_o is a regular point, Theorem 15.1 supplies V harmonic and > 0 in Ω with $V(x) \to 0$ as $x \to x_o$. The proof of Theorem 15.1 shows that V can be taken to be bounded in Ω. By Theorem 11.2, V extends to be harmonic in $\dot{B}$. But $V(x_o) = 0$ and $V \geq 0$ on $\dot{B}$, and $V > 0$ on $\dot{B}\setminus F$. This is impossible.

Thus every point of F is irregular for Ω.

We now shall give a quantitative condition which assures that a given point on $\partial\Omega$ is regular.

Theorem 16.2: Let Ω be a bounded region in $\mathbb{R}^3$ and $x_o \in \partial\Omega$. Assume $\exists\, k > 0$ such that for all $r > 0$ sufficiently small

$$C((\mathbb{R}^3\setminus\Omega) \cap \{x \mid |x-x_o| \leq r\}) > k\cdot r. \tag{16.7}$$

Then x_o is a regular point for Ω.

Note: The hypothesis (16.7) assures that the complement of Ω is sufficiently fat near x_o.

Proof: We proceed, as in the last theorem, with $0 \in \Omega$ and $\overline{\Omega} \subset \{x \mid |x| < 1\}$. As we saw in the proof of the last theorem, to have x_o regular for Ω it suffices to have $Tx_o \in E\setminus\tilde{E}$. Here $E = \mathbb{R}^3\setminus T(\Omega\setminus\{0\})$, as above. Put $x_o' = Tx_o$. In order that $x_o' \in E\setminus\tilde{E}$, it suffices by Theorem 10.1 that

$$C(E \cap \{x \mid |x-x_o'| \leq \rho\} > c_o\cdot\rho \tag{16.8}$$

for all small ρ, where c_o is a constant.

Fix a neighborhood of $\partial E \cup \partial\Omega$ and a constant c_1 such that

$$|Ta-Tb| \leq c_1|a-b| \tag{16.9}$$

for a,b in the neighborhood. Fix $\rho > 0$. Put $B(\overline{x},s) = \{y \mid |y-\overline{x}| \leq s\}$. It is easily seen that

(16.10) $$B(x_o, \frac{1}{c_1} \cdot \rho) \subseteq T(B(x_o', \rho)).$$

Hence

$$B(x_o, \frac{\rho}{c_1}) \cap (\mathbb{R}^3 \setminus \Omega) \subseteq T\{B(x_o', \rho) \cap E\}.$$

So, by (16.7), for small ρ

$$C(T\{B(x_o', \rho) \cap E\}) \geq k \cdot \frac{\rho}{c_1}$$

By Exercise 16.2, $\exists M$, independent of ρ, such that

$$C(T\{B(x_o', \rho) \cap E\}) \leq M \cdot C\{B(x_o', \rho) \cap E\}.$$

Hence for small ρ,

$$C\{B(x_o', \rho) \cap E\} \geq \frac{k}{Mc_1} \cdot \rho;$$

hence (16.8) holds, and we are done.

Corollary: Let Ω be a bounded region in $\mathbb{R}^3$ and $x_o \in \partial\Omega$. Assume $\exists$ a cone K with vertex at x_o such that for some r_o, $\mathbb{R}^3 \setminus \Omega$ contains $K \cap \{x | \ |x-x_o| \leq r_o\}$. Then x_o is a regular point for Ω.

Proof: Put $K^{(r)} = K \cap \{x | \ |x-x_o| \leq r\}$. We saw in Section 10 that for some constant c_1,

$$C(K^{(r)}) \geq c_1 \cdot r.$$

Then for $r \leq r_o$,

$$C\{(\mathbb{R}^3 \setminus \Omega) \cap (|x-x_o| \leq r)\} \geq c_1 r,$$

i.e. (16.7) holds. The Theorem thus gives the assertion.

<u>Note</u>: Corollary 1 of Theorem 15.2 is contained in the last result.

17. The Riesz Decomposition Theorem

In Section 15 we defined a function to be superharmonic if its negative is subharmonic. This is equivalent to saying that F is superharmonic on the open set Ω in $\mathbb{R}^n$ if for every $x_0 \in \Omega \; \exists \; r(x_0) > 0$ with

$$F(x_0) \geq \frac{1}{A} \int_{|x-x_0|=a} F(x)dS, \tag{17.1}$$

for all $a < r(x_0)$, where A is the area of the sphere $\{x \mid |x-x_0| = a\}$, and

$$\varliminf_{x \to x_0} F(x) \geq F(x_0), \quad \text{for all } x_o \in \Omega. \tag{17.2}$$

We allow F to equal $+\infty$ at a point, but not $-\infty$. Also F is not identically $+\infty$ on any ball.

Obviously, every harmonic function is superharmonic (as well as subharmonic). What other superharmonic functions are there?

Example: $F(x) = \frac{1}{|x|}$, defined at 0 by $F(0) = \infty$, is superharmonic in $\mathbb{R}^3$, since (17.1), (17.2) are satisfied at every point of $\mathbb{R}^3$.

Theorem 17.1: Let μ be a measure of compact support on $\mathbb{R}^n$. Then U^μ is superharmonic in $\mathbb{R}^n$.

Proof: Recall that in Section 5 we saw that

$$\int_{|y|=a} \frac{dS_y}{|x-y|} = \begin{cases} \frac{4\pi a^2}{|x|}, & |x| > a \\ \frac{4\pi a^2}{a}, & |x| \leq a \end{cases} . \tag{17.3}$$

Fix $x_0 \in \mathbb{R}^3$ and $a > 0$.

$$\int_{|x-x_0|=a} U^\mu(x)\,dS = \int d\mu(\zeta) \left\{ \int_{|x-x_0|=a} \frac{dS_x}{|x-\zeta|} \right\}.$$

$$\int_{|x-x_0|=a} \frac{dS_x}{|x-\zeta|} = \int_{|y|=a} \frac{dS_y}{|x_0+y-\zeta|} = \int_{|y|=a} \frac{dS_y}{|(\zeta-x_0)-y|} \le \frac{4\pi a^2}{|\zeta-x_0|}$$

because of (17.3). Hence

$$\int_{|x-x_0|=a} U^\mu(x)\,dS \le 4\pi a^2 \int \frac{d\mu(\zeta)}{|\zeta-x_0|},$$

or

$$\frac{1}{4\pi a^2} \int_{|x-x_0|=a} U^\mu(x)\,dS \le U^\mu(x_0).$$

Thus U^μ satisfies (17.1). That (17.2) holds for U^μ we know by Lemma 7.5. Thus the Theorem is proved.

Since the sum of two superharmonic functions evidently is superharmonic, we have: if Ω is a region in $\mathbb{R}^n$, and if h is harmonic in Ω and μ is a measure of compact support, then

$$U^\mu + h$$

is superharmonic in Ω.

In 1930, Fredrick Riesz showed that this is essentially the form of the most general superharmonic function.

We shall prove a result of this type where the region in question is a ball and the function to be represented satisfies a mild restriction.

Theorem 17.2: (F. Riesz) Let B be a closed ball, V a function superharmonic in a region containing B, where V is continuous outside of $\dot{B}$. Then $\exists$ measure μ on B and a function H harmonic in $\dot{B}$ such that

$$V(x) = U^{\mu}(x) + H(x), \quad x \in \dot{B}. \tag{17.4}$$

Note 1: When V is smooth (17.4) is an easy consequence of things we know.

Let $V \in C^2$ in a region containing B. Let ρ denote the restriction to $\dot{B}$ of $-\frac{1}{4\pi}\Delta V$. Because of Exercise 14.2, $\Delta V \leq 0$ and so $\rho \geq 0$. Hence $\mu = \rho dx$ is a measure.

Theorem 3.5 gives

$$\Delta U^{\mu} = -4\pi\rho = \Delta V \quad \text{in} \quad \dot{B}.$$

Hence $\Delta(V-U^{\mu}) = 0$ in $\dot{B}$, so

$$H = V - U^{\mu}$$

is harmonic in $\dot{B}$, and (17.4) holds.

To prove Theorem 17.2 we shall reduce the problem to smooth superharmonic functions by suitably approximating arbitrary superharmonic functions by smooth ones.

Note 2: In proving Theorem 17.2 we may assume without loss of generality that V is harmonic outside of B in some neighborhood of B. For we can find V_1 superharmonic in a neighborhood of B, harmonic outside of B, and coinciding with V in B.

We obtain V_1 as follows: let B_1 be a closed ball containing B in its interior with V continuous on $B_1 \setminus \dot{B}$ and superharmonic on B_1. In $B_1 \setminus \dot{B}$ we take V_1 to be the harmonic function which agrees with V on $\partial B_1 \cup \partial B$. In $\dot{B}$, we put $V_1 = V$.

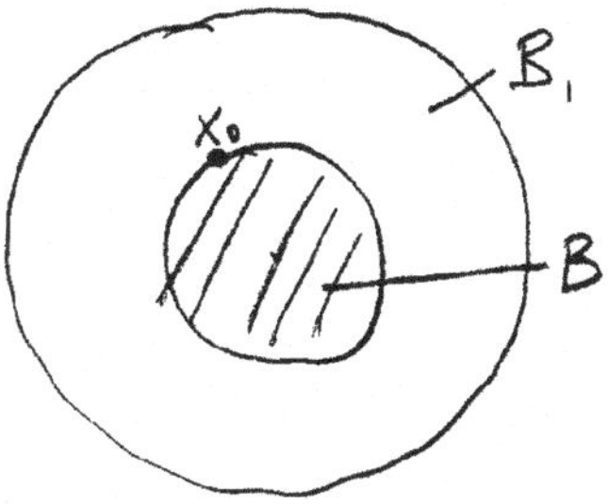

V_1 evidently satisfies (17.1) at all points of $\dot{B}_1$ except possibly on ∂B.

Fix $x_o \in \partial B$. For small r

$$V(x_o) \geq \frac{1}{4\pi r^2} \int_{|x-x_o|=r} V dS. \tag{17.5}$$

Since $V_1 = V$ on the boundary of $\dot{B}_1 \setminus B$, $V \geq V_1$ on $\dot{B}_1 \setminus B$ and so $V \geq V_1$ on $\{x \mid |x-x_o| = r\}$. (17.5) now gives

$$V_1(x_o) = V(x_o) \geq \frac{1}{4\pi r^2} \int_{|x-x_o|=r} V_1 dS,$$

i.e., V_1 satisfies (17.1) at x_o. Also V_1 satisfies (17.2) at each point of B_1, so V_1 has the desired properties.

Lemma 17.3: *Let* W *be a bounded region in* $\mathbb{R}^n$ *and let* v *be superharmonic in some neighborhood of* $\overline{W}$. *Then* $\exists$ *sequence* $\{v_j\}$ *of functions defined on* W *such that*

(i) *Each* $v_j \in C^\infty(W)$.

(ii) *Each* v_j *is superharmonic in* W.

(iii) $v_j(x) \leq v(x)$ *for all* j, $x \in W$ *and* $\lim_{j \to \infty} v_j(x) = v(x)$ *for all* $x \in W$.

(iv) *If* v *is harmonic on some neighborhood of a compact set* K, *then* $v_j = v$ *on* K *for all large* j.

(v) $\exists$ constant C such that $C \leq v_j$ on W for all j.

We postpone the proof of the Lemma.

Proof of Theorem 17.2. By Note 2 above, we may assume without loss of generality that V is superharmonic on a neighborhood of some ball B_1 with $B \subset \dot{B}_1$, and V is harmonic outside B.

Choose $\{v_j\}$ by Lemma 17.3 for $v = V$ on a neighborhood of B_1.

Assertion 1: $\exists$ constant M such that

$$\left| \int_B \Delta f \cdot V dx \right| \leq M \cdot \max_B |f|, \tag{17.6}$$

for all $f \in C_0^2(\dot{B})$.

It follows from (17.2) that V is bounded below on B_1, and from this, together with (17.1), we know that $\int_B |V| dx < \infty$.

For each j, Green's formula yields whenever $f \in C_0^2(\dot{B})$

$$\int_B \Delta f \cdot v_j dx = \int_B f \cdot \Delta v_j dx. \tag{17.7}$$

Also by Green's formula

$$\int_{B_1} \Delta v_j dx = \int_{\partial B_1} \frac{\partial v_j}{\partial n} dS.$$

By (iv) of the Lemma, and the fact that V is harmonic outside B, $v_j = V$ in some neighborhood of ∂B_1, for all large j. Hence

$$\frac{\partial v_j}{\partial n} = \frac{\partial V}{\partial n} \quad \text{on } \partial B_1$$

for all large j. Thus we have

$$\int_{B_1} \Delta v_j dx = \int_{\partial B_1} \frac{\partial V}{\partial n} ds. \tag{17.8}$$

Since v_j is superharmonic, $\Delta v_j \leq 0$, so

$$\int_B |\Delta v_j| dx \leq \int_{B_1} |\Delta v_j| dx = -\int_{B_1} \Delta v_j dx.$$

(17.8) then gives that $\exists M$ with

$$\int_B |\Delta v_j| dx \leq M, \text{ all } j.$$

Applying this to (17.7), we get

$$\left|\int_B \Delta f \cdot v_j dx\right| \leq M \cdot \max_B |f| \tag{17.9}$$

for all j.

By the Lemma, $C \leq v_j \leq V$ on B and $v_j \to V$ pointwise. Since V is summable over B, by the dominated convergence theorem (17.9) gives (17.6), proving Assertion 1.

Let L denote the linear map: $C_0^2(\dot{B}) \to \mathbb{R}$ given by:

$$L(f) = -\int_B \Delta f \cdot V dx.$$

In view of (17.6), L is a bounded linear functional, where $\|f\| = \max_B |f|$. The Riesz representation theorem for linear functionals supplies us with a signed measure μ on B such that $L(f) = \int f d\mu$, i.e.,

(17.10) $$-\int_B \Delta f\cdot V dx = \int_B f d\mu, \quad f \in C_0^2(\dot{B}).$$

We may evidently assume that μ vanishes identically on ∂B.

We claim $\mu \geq 0$. For if $f \geq 0$,

$$-\int_B \Delta f v_j dx \geq 0$$

by (17.7), since $\Delta v_j \leq 0$. Letting $j \to \infty$, we get

$$-\int_B \Delta f\cdot V dx \geq 0$$

whenever $f \geq 0$ and $f \in C_0^2(\dot{B})$. Hence $\int_B f d\mu \geq 0$ for all such f, and so $\mu \geq 0$, as claimed.

Recall the notion of distribution, discussed in Section 4.

Definition 17.1: Let Φ_1, Φ_2 be two distributions in $\mathbb{R}^n$ and W a domain in $\mathbb{R}^n$ We say

$$\Phi_1 = \Phi_2 \quad \text{on } W,$$

provided $\Phi_1(f) = \Phi_2(f)$ for all $f \in C_0^\infty(W)$.

Let us denote by $\tilde{V}$ the distribution on $\mathbb{R}^3$ given by

$$\tilde{V}(f) = \int_B fV dx, \quad f \in C_0^\infty(\mathbb{R}^3),$$

and write μ for the distribution on $\mathbb{R}^3$ given by μ. By definition of the derivative of a distribution,

$$\Delta\tilde{V}(f) = \int_B V\cdot\Delta f dx, \quad f \in C_0^\infty.$$

Because of (17.10)

$$\Delta\tilde{V}(f) = -\mu(f), \quad f \in C_o^2(\dot{B}),$$

so by Definition 17.1,

$$\Delta\tilde{V} = -\mu \quad \text{on} \quad \dot{B}.$$

By Theorem 4.1 on the other hand

$$\Delta U^\mu = -4\pi\mu.$$

Thus

$$\Delta(\tilde{V} - \frac{1}{4\pi} U^\mu) = 0 \quad \text{on} \quad \dot{B}. \tag{17.11}$$

We now appeal to

Proposition 17.4: (Weyl's Lemma) Let F be a locally summable function on $\mathbb{R}^n$ and W a bounded region in $\mathbb{R}^n$ such that

$$\Delta\tilde{F} = 0 \quad \text{in} \quad W. \tag{17.12}$$

Then F is a harmonic function in W, i.e., $\exists$ a harmonic function G such that $F = G$ a.e. in W.

Note: Of course, if we know a priori that $F \in C^2$ in W, there is nothing to prove. The content of the Proposition is that (17.12) implies that F is (essentially) smooth in W.

We shall prove this result below. By Weyl's Lemma, $V - \frac{1}{4\pi} U^\mu$ is harmonic in

$\dot{B}$, i.e. $\exists$ H harmonic in $\dot{B}$ such that

$$V - \frac{1}{4\pi} U^{\mu} = H \text{ a.e. in } \dot{B}.$$

It follows that for $x_0 \in \dot{B}$ and almost all small ϵ,

$$V - \frac{1}{4\pi} U^{\mu} = H \text{ a.e. } -dS \text{ on } |x-x_o| = \epsilon. \tag{17.13}$$

Exercise 17.1: If φ is superharmonic in a region in $\mathbb{R}^3$ then for each x_0

$$\lim_{\epsilon \to 0} \frac{1}{4\pi\epsilon^2} \int_{|x-x_0|=\epsilon} \varphi dS = \varphi(x_0).$$

Applying the Exercise to (17.13) gives that $V - \frac{1}{4\pi} U^{\mu} = H$ at x_0, for every $x_0 \in \dot{B}$, i.e., $V = U^{\frac{1}{4\pi}\mu} + H$ on $\dot{B}$, proving (17.4). q.e.d.

We have to supply proofs for Lemma 17.3 and Proposition 17.4.

Let Ω be a bounded region in $\mathbb{R}^3$. Let Ω_1 be a region $\subset \Omega$ and $\epsilon_1 > 0$ such that if $\bar{x} \in \Omega_1$, $\{x \mid |x-\bar{x}| \leq \epsilon_1\} \subset \Omega$.

Fix $\chi \in C^{\infty}(\mathbb{R}^3)$, $\chi \geq 0$ and $\chi(y) = 0$ for $|y| \geq 1$, such that χ is a function of $|y|$, and $\int \chi(y)dy = 1$.

Definition 17.2: Let φ be a summable function defined on Ω.

$$\varphi_{\epsilon}(x) = \int \varphi(x-\epsilon y)\chi(y)dy \tag{17.14}$$

where $0 < \epsilon \leq \epsilon_1$.

If $x \in \Omega_1$, $x - \epsilon y \in \Omega$ for $|y| \leq 1$, so φ_{ϵ} is defined on Ω_1.

Lemma 17.5:

$$\lim_{\epsilon \to 0} \int_{\Omega_1} |\varphi_\epsilon(x) - \varphi(x)|\,dx = 0.$$

Proof:

$$\varphi_\epsilon(x) - \varphi(x) = \int [\varphi(x-\epsilon y) - \varphi(x)]\chi(y)dy,$$

so

$$|\varphi_\epsilon(x) - \varphi(x)| \le \int |\varphi(x-\epsilon y) - \varphi(x)|\chi(y)dy.$$

Hence

$$\int_{\Omega_1} |\varphi_\epsilon(x) - \varphi(x)|\,dx \le \int \chi(y)dy \int_{\Omega_1} |\varphi(x-\epsilon y) - \varphi(x)|\,dx. \tag{17.15}$$

Since

$$\int_\Omega |\varphi|\,dx < \infty,$$

$$\lim_{\epsilon \to 0} \int_{\Omega_1} |\varphi(x-\epsilon y) - \varphi(x)|\,dx = 0$$

for each y in $|y| \le 1$. Also

$$\int_{\Omega_1} |\varphi(x-\epsilon y) - \varphi(x)|\,dx \le 2 \int_\Omega |\varphi(x)|\,dx,$$

for each y in $|y| \le 1$. Hence the integral on the right in (17.15) converges to 0 as $\epsilon \to 0$, by dominated convergence.

(17.15) thus gives the assertion.

Exercise 17.2: For each ϵ, $\varphi_\epsilon \in C^\infty(\Omega_1)$.

Lemma 17.6: Let φ be summable on Ω and such that $\widetilde{\Delta\varphi} = 0$ in Ω. Let Ω_1, ϵ_1 be as above. Then φ_ϵ is harmonic in Ω_1 for each $\epsilon \leq \epsilon_1$.

Proof: Fix $f \in C_o^2(\Omega_1)$

$$\begin{aligned}\int f\Delta\varphi_\epsilon dx &= \int \Delta f\cdot\varphi_\epsilon dx \\ &= \int \Delta f(x)dx \int \varphi(x-\epsilon y)\chi(y)dy \\ &= \int \chi(y)dy \int \varphi(x-\epsilon y)\Delta f(x)dx \\ &= \int \chi(y)dy \int \varphi(\zeta)\Delta f(\zeta+\epsilon y)d\zeta.\end{aligned}$$

But

$$\Delta f(\zeta+\epsilon y) = \Delta_\zeta\{(\zeta+\epsilon y)\},$$

so

$$\int \varphi(\zeta)\Delta f(\zeta+\epsilon y)d\zeta = \int \varphi(\zeta)\Delta_\zeta\{f(\zeta+\epsilon y)\}d\zeta.$$

Since $f(\zeta+\epsilon y) \in C_o^2(\Omega)$, and since $\widetilde{\Delta\varphi} = 0$ in Ω, by hypothesis, the last integral on the right $= 0$. So

$$\int f\Delta\varphi_\epsilon dx = 0, \quad \text{all } f \in C_o^2(\Omega_1).$$

Hence $\Delta\varphi_\epsilon = 0$ in Ω_1. q.e.d.

Lemma 17.7: With assumptions as in the last Lemma, let B be a closed ball $\subset \Omega_1$. Then, $\{\varphi_\epsilon | \epsilon > 0\}$ is uniformly bounded in $\dot{B}$.

Proof: Since $B \subset \Omega_1$, $\exists\, \delta > 0$ such that whenever $x_o \in B$, $\{x | |x-x_o| \leq \delta\} \subset \Omega_1$. Fix ϵ. Since φ_ϵ is harmonic in Ω_1, if $x_o \in B$,

$$\varphi_\epsilon(x_0) = \frac{1}{V_\delta} \int_{|x-x_0|\leq\delta} \varphi_\epsilon(x)dx,$$

where V_δ is the volume of the ball $|x-x_0| \leq \delta$, and so

$$|\varphi_\epsilon(x_0)| \leq \frac{1}{V_\delta} \int_{\Omega_1} |\varphi_\epsilon(x)|\, dx \leq \frac{1}{V_\delta} \int_\Omega |\varphi|\, dx,$$

whence we have the desired bound.

Proof of Proposition 17.4: Let W_1 be a region with $\overline{W}_1 \subset W$. For all small ϵ, F_ϵ is defined by Definition 17.2 in W_1.

$\Delta\tilde{F} = 0$ in W, by hypothesis, so F_ϵ is harmonic in W_1 for all small ϵ.

Let B be a closed ball in W_1. By Lemma 17.7, $\{F_\epsilon\}$ is uniformly bounded in $\dot{B}$ as $\epsilon \to 0$. Hence $\{F_\epsilon\}$ has a subsequence $\{F_{\epsilon_j}\}$ converging boundedly on $\dot{B}$ to a harmonic function H. Hence

$$\int_B |F_{\epsilon_j} - H|\, dx \to 0 \quad \text{as} \quad j \to \infty.$$

Also

$$\int_B |F_\epsilon - F|\, dx \to 0 \quad \text{as} \quad \epsilon \to 0,$$

by Lemma 17.5.

Hence $F = H$ a.e. in B. Thus for each ball $B \subset W$, we can find la harmonic function in $\dot{B}$ which $= F$ a.e. in B. Hence F is harmonic in W.

q.e.d.

Note: When φ is superharmonic, the set $\{\varphi_\epsilon\}$ has particularly nice properties, which will give us Lemma 17.3.

Let now Ω, Ω_1, X and φ_ϵ be defined as in Definition 17.2.

Lemma 17.8: Assume that φ is superharmonic in Ω. Then

(17.16) $$\varphi_\epsilon(x) \leq \varphi(x) \quad \text{for all} \quad \epsilon, x \in \Omega_1.$$

(17.17) $$\lim_{\epsilon \to 0} \varphi_\epsilon(x) = \varphi(x), \quad x \in \Omega_1.$$

(17.18) $$\exists \text{ constant } C \text{ such that } C \leq \varphi_\epsilon \text{ on } \Omega_1 \text{ for all } \epsilon.$$

(17.19) For each ϵ, φ_ϵ is superharmonic in Ω_1.

(17.20) If K is a compact subset of Ω such that φ is harmonic in a neighborhood of K, then $\varphi_c = \varphi$ on K, for all small ϵ.

Proof: Recall that

$$\varphi_\epsilon(x) = \int \varphi(x-\epsilon y)\chi(y)dy,$$

where $\chi \geq 0$, $\chi \in C^\infty$, χ is a function of $|y|$, i.e., $\chi(y) = \chi(|y|)$, χ vanishes for $|y| > 1$, and $\int \chi(y)dy = 1$.

We note the formula

(17.21) $$\int_{|y|\leq 1} f(y)dy = \int_0^1 r^2 dr \int_{|\zeta|=1} f(r\zeta)dS_\zeta,$$

where f is a summable function.

The formula gives

(17.22) $$\varphi_\epsilon(x) = \int_0^1 \chi(r)r^2 dr \int_{|\zeta|=1} \varphi(x-\epsilon r\zeta)dS_\zeta.$$

Also

(17.23) $$\int_{|\zeta|=1} \varphi(x-\epsilon r\zeta)dS_\zeta = \frac{1}{r^2\epsilon^2} \int_{|x-y|=r\epsilon} \varphi(y)dS_y.$$

Since φ is superharmonic, the right side $\leq 4\pi\varphi(x)$. (17.22) thus gives

$$\varphi_\epsilon(x) \leq 4\pi\varphi(x) \int_0^1 \chi(r)r^2 dr = \varphi(x),$$

by (17.21) and the fact that $\int \chi(y)dy = 1$. Thus (17.16) holds.

$$\lim_{\epsilon \to 0} \int_{|\zeta|=1} \varphi(x-\epsilon r\zeta)dS_\zeta = 4\pi \lim_{\rho \to 0} \left\{ \frac{1}{4\pi\rho^2} \int_{|x-y|=\rho} \varphi(y)dS_y \right\},$$

because of (17.23).

The term on the right $= 4\pi\varphi(x)$, by Exercise 17.1. Thus (17.22) yields that

$$\lim_{\epsilon \to 0} \varphi_\epsilon(x) = \int_0^1 \chi(r)r^2 dr \cdot 4\pi\varphi(x),$$

by dominated convergence, and the right side $= \varphi(x)$. So (17.17) holds.

Since φ is superharmonic, $\exists$ constant C with $\varphi \geq C$ on a neighborhood of Ω_1. Hence for $x \in \Omega_1$,

$$\varphi_\epsilon(x) \geq \int C\chi(y)dy = C,$$

so (17.18) holds.

Let next K be as in (17.20). Choose ϵ so that for all $x \in K$, the ball of radius ϵ around x lies in the region where φ is harmonic. Then for $x \in K$,

$$\varphi_\epsilon(x) = 4\pi \int_0^1 \chi(r)r^2 dr \left\{ \frac{1}{4\pi r^2 \epsilon^2} \int_{|x-y|=r\epsilon} \varphi(y)dS_y \right\}.$$

The expression in braces $= \varphi(x)$. So

$$\varphi_\epsilon(x) = 4\pi \int_0^1 \chi(r) r^2 dr \cdot \varphi(x) = \varphi(x),$$

proving (17.20).

The verification of (17.19) is left to the reader. The Lemma is established.

Proof of Lemma 17.3: Put $\varphi = v$. Choose a sequence $\{\epsilon_j | j = 1,2,\ldots\} \to 0$ and put $v_j = \varphi_{\epsilon_j}$ for all j. Because of the properties of the φ_ϵ proved in Lemma 17.8, $\{v_j\}$ satisfies all the assertions of Lemma 17.3.

18. Applications of the Riesz Decomposition

Let V be a function defined on $\mathbb{R}^n$. When is V a potential, i.e. when can we find a measure μ of compact support such that $V = U^{\mu}$?

The following conditions are evidently necessary:

(18.1) $$V > 0 \quad \text{everywhere.}$$

(18.2) $$V(x) \to 0 \quad \text{as} \quad x \to \infty.$$

(18.3) $$V \text{ is superharmonic in } \mathbb{R}^n.$$

(18.4) $$V \text{ is harmonic outside some ball.}$$

<u>Theorem 18.1</u>: <u>Conditions (18.1) through (18.4) are sufficient to assure that V is a potential.</u>

<u>Proof</u>: Fix R_0, by (18.4), such that V is harmonic for $|x| > R_0 - 1$.

By Theorem 17.2, $\exists$ measure μ_0 on $|x| \leq R_0$ and H_0 harmonic in $|x| < R_0$ such that

(18.5) $$V = U^{\mu_0} + H_0 \quad \text{in} \quad |x| < R_0.$$

Choose $R > R_0$. In the same way we get a measure μ_R on $|x| \leq R$ and H_R harmonic in $|x| < R$ such that

(18.6) $$V = U^{\mu_R} + H_R \quad \text{in} \quad |x| < R.$$

<u>Assertion</u>: $\mu_R = \mu_0$.

For by (18.6),

$$0 = \Delta\tilde{V} = -4\pi\mu_R \quad \text{in} \quad R_0 - 1 < |x| < R,$$

while by (18.5) and (18.6),

$$-4\pi\mu_R = \Delta\tilde{V} = -4\pi\mu_0 \quad \text{on} \quad |x| < R_0,$$

so

$$\mu_R = \mu_0 \quad \text{on} \quad |x| < R_0.$$

Hence μ_R and μ_0 coincide as measures on $\mathbb{R}^3$. (18.6) thus gives

$$V - U^{\mu_0} = H_R \quad \text{in} \quad |x| < R.$$

Since R was arbitrary, we conclude that if $H = V - U^{\mu_0}$, then H is harmonic in $\mathbb{R}^3$. Now for large $|x|$, given ϵ, $|V| < \epsilon$, by (18.2) and $U^{\mu_0} < \epsilon$, so

$$|H| \leq 2\epsilon \quad \text{for} \quad |x| = t, \quad t \text{ large}.$$

By the maximum principle,

$$|H| < 2\epsilon \quad \text{for} \quad |x| \leq t.$$

Since ϵ is arbitrary, $H = 0$, and

$$V = U^{\mu_0}, \quad \text{q.e.d.}$$

Problem: What functions are there that are positive and harmonic in a punctured ball $0 < |x| < R$ in $\mathbb{R}^3$?

The ones that come to mind are, first, functions positive and harmonic in the entire ball, and, second, $\frac{c}{|x|}$ where c is a positive constant.

Theorem 18.2: (Bôcher) If U is positive and harmonic in a punctured ball $\dot{B}\setminus\{0\}$ in $\mathbb{R}^3$, then $\exists$ a non-negative constant b and H harmonic in $\dot{B}$ with

$$U(x) = \frac{b}{|x|} + H(x).$$

Proof: Define

$$V(x) = U(x) + \frac{1}{|x|} \quad \text{for } x \neq 0$$

$$V(0) = \infty.$$

We verify at once that V is superharmonic in $\dot{B}$. The positivity of U is needed in order that $\lim\limits_{x \to 0} V(x) = V(0)$. Choose a ball $B_2 : |x| \leq R$ with $B_2 \subset \dot{B}$.

By Theorem 17.2, $\exists$ measure μ on B_2 and H harmonic in $\dot{B}_2$ with

$$V = U^{\mu} + H, \quad \text{in } \dot{B}_2.$$

Then

$$\Delta V = \Delta(U + \frac{1}{|x|}) = -4\pi\mu,$$

so $\mu = 0$ in $0 < |x|$, whence μ is a point mass at 0. Thus

$$V(x) = \frac{c}{|x|} + H(x),$$

so

$$U(x) = \frac{c-1}{|x|} + H(x).$$

If $c - 1 < 0$, U would assume negative values near $x = 0$. Hence $c - 1 \geq 0$. q.e.d.

Corollary 1: Let φ be positive and harmonic in $|y| > R$ for some R. Then $\varphi(y)$ has a limit as $y \to \infty$.

Proof: Put $V(x') = |x'|\varphi(x')$, and

$$U(x) = V(Tx).$$

Since $\frac{V(x')}{|x'|} = \varphi(x')$ is harmonic in $|x'| > R$, $U(x)$ is harmonic for $0 < |x| < \frac{1}{R}$. Also $U > 0$, since $\varphi > 0$. By the preceding theorem,

$$U(x) = \frac{b}{|x|} + H(x),$$

where H is harmonic for $|x| < \frac{1}{R}$.

$$\varphi(x') = \frac{U(Tx')}{|x'|} = b + \frac{1}{|x'|} H(Tx').$$

As $x' \to \infty$, $Tx' \to 0$, $H(Tx')$ remains bounded, and so $\varphi(x') \to b$. q.e.d.

Corollary 2: Let φ be positive and harmonic in all of $\mathbb{R}^3$. Then φ is a constant.

The proof is left as an Exercise. We can now generalize Bôcher's theorem by replacing the singular set consisting of the point 0, by a general compact subset of $\dot{B}$ of capacity 0.

Theorem 18.3: Let B be a closed ball in $\mathbb{R}^3$, E a compact subset of $\dot{B}$ with $C(E) = 0$. Let V be a function positive and harmonic on $\dot{B}\setminus E$. Then $\exists$ measure σ on E and H harmonic in $\dot{B}$ such that

$$V = U^\sigma + H \text{ in } \dot{B}\setminus E. \tag{18.7}$$

We need the following result, which is of independent interest.

Proposition 18.4: Let E be a compact set in $\mathbb{R}^3$ with $C(E) = 0$. Then $\exists$ a measure μ of compact support in $\mathbb{R}^3$ such that $U^\mu = \infty$ on E, $U^\mu < \infty$ on $\mathbb{R}^3 \setminus E$.

Proof: Choose smoothly bounded open sets $E_1, E_2, \ldots$ with

$$E \subset E_j, \quad \overline{E}_{j+1} \subset E_j, \text{ all } j$$

and

$$E = \bigcap_{j=1}^{\infty} E_j .$$

Since $C(E) = 0$, $\lim_{j \to \infty} C(E_j) = 0$, by Lemma 11.1. Without loss of generality

$$C(E_n) \leq \frac{1}{n} \cdot \frac{1}{2^n}, \quad n = 1,2,\ldots .$$

Let μ_n be the equilibrium distribution on E_n.

$$U^{\mu_n} = \frac{1}{C(E_n)} \geq n \cdot 2^n \quad \text{on } E_n .$$

Put

$$\overline{\mu} = \frac{1}{2}\mu_1 + \frac{1}{2^2}\mu_2 + \frac{1}{2^3}\mu_3 + \cdots .$$

The series converges to a measure $\overline{\mu}$ supported on $\overline{E}_1$.

$$U^{\overline{\mu}} = \frac{1}{2}U^{\mu_1} + \frac{1}{2^2}U^{\mu_2} + \cdots .$$

Since $U^{\mu_n} \geq n \cdot 2^n$ on E_n and hence on E,

$$U^{\mu} \geq \frac{1}{2^n} U^{\mu_n} \geq n \quad \text{on } E, \text{ for each } n.$$

Hence $U^{\overline{\mu}} = \infty$ on E.

Fix $x \in \mathbb{R}^3 \setminus E$. Choose n_x so that $x \notin \overline{E}_{n_x}$ and let dx be the distance from x to $\overline{E}_{n_x}$. Then $\operatorname{dist}(x, E_n) \geq dx$ for all $n \geq n_x$. Hence

$$U^{\mu_n}(x) \leq \frac{1}{dx} \quad \text{if} \quad n \geq n_x,$$

and so

$$U^{\bar{\mu}}(x) \leq \sum_{n=1}^{n_x} \frac{1}{2^n} U^{\mu_n}(x) + \sum_{n=n_x+1}^{\infty} \frac{1}{2^n} \cdot \frac{1}{dx} .$$

Since each U^{μ_n} is bounded, $U^{\bar{\mu}}(x) < \infty$. Thus $U^{\bar{\mu}} < \infty$ on $\mathbb{R}^3 \setminus E$, and we are done.

Lemma 18.5: *Let φ be positive superharmonic in a region, B a closed ball in the region, $x_0 \in \dot{B}$. Then*

$$\varphi(x_0) \geq -\frac{1}{4\pi} \int_{\partial B} \varphi \frac{\partial G_{x_0}}{\partial n} dS,$$

where G_{x_0} is the Green's function of $\dot{B}$.

Proof: Form φ_ϵ by Definition 17.2. φ_ϵ is smooth and superharmonic in a neighborhood of B.

Let Φ_ϵ be the function harmonic in $\dot{B}$ which agrees with φ_ϵ on ∂B. Then

$$\varphi_\epsilon(x_0) \geq \Phi_\epsilon(x_0) = -\frac{1}{4\pi} \int_{\partial B} \varphi_\epsilon \frac{\partial G_{x_0}}{\partial n} dS. \tag{18.8}$$

By (17.17)

$$\lim_{\epsilon \to 0} \varphi_\epsilon(x_0) = \varphi(x_0)$$

and

$$\lim_{\epsilon \to 0} \varphi_\epsilon(x) = \varphi(x)$$

for $x \in \partial B$. Also by (17.16)

$$\varphi_\epsilon(x) \leq \varphi(x) \quad \text{for} \quad x \in \partial B.$$

Also $0 \leq \varphi_\epsilon$ since $\varphi > 0$ by assumption. Hence

$$\int_{\partial B} \varphi_\epsilon \frac{\partial G_{x_o}}{\partial n} \, dS \to \int_{\partial B} \varphi \frac{\partial G_{x_o}}{\partial n} \, dS$$

by dominated convergence. So letting $\epsilon \to 0$ in (18.8), we get the assertion of the Lemma.

Proof of Theorem 18.3: By Proposition 18.4, we can choose a measure μ such that $U^\mu = \infty$ on E, $U^\mu < \infty$ outside of E. Put

$V_t = V + t \cdot U^\mu$, t a positive scalar. Define $V_t = \infty$ on E. Fix $a \in E$.

Since $\underline{\lim}_{x \to a} U^\mu(x) = \infty$ and $V > 0$, $\lim_{x \to a} V_t(x) = \infty$. Hence V_t is superharmonic in B.

Fix $a \in E$ and consider a ball D: $\{x \mid |x-a| \leq r\}$. $C(E \cap \partial D) \leq C(E) = 0$. It follows that the area of $E \cap \partial D$ on $\partial D = 0$. (Why?). Fix $x_o \in \dot{D} \setminus E$.

Fix t. By Lemma 18.5

$$V_t(x_o) \geq -\frac{1}{4\pi} \int_{\partial D} V_t \frac{\partial G_{x_o}}{\partial n} \, dS.$$

Since $E \cap \partial D$ has zero area, the integral can be taken over $\partial D \setminus E$. For $x \in \partial D \setminus E$, $V_t(x) = V(x) + tU^\mu(x)$ decreases toward $V(x)$ as $t \downarrow 0$. Also $V_t(x_o) \downarrow V(x_o)$. Hence by monotone convergence, we conclude

(18.9) $$V(x_o) \geq \frac{-1}{4\pi} \int_{\partial D\setminus E} V \frac{\partial G_{x_o}}{\partial n} \, ds.$$

As $x_o \to a$, $-\frac{1}{4\pi}\frac{\partial G_{x_o}}{\partial n} \to -\frac{1}{4\pi}\frac{\partial G_a}{\partial n} = \frac{1}{4\pi r^2}$. (18.9) thus yields

(18.10) $$\underline{\lim}_{x_o \to a} V(x_o) \geq \frac{1}{4\pi r^2} \int_{\partial D\setminus E} V ds.$$

We now extend the definition of V to all of $\dot{B}$ by setting for $a \in E$:

$$V^*(a) = \underline{\lim}_{x \to a} V(x)$$

and putting $V^* = V$ on $\dot{B}\setminus E$.

We claim V^* is superharmonic on $\dot{B}$. It follows from the definition that V^* satisfies (17.2) at each point. (18.10) gives for $a \in E$, for small r,

$$V^*(a) \geq \frac{1}{4\pi r^2} \int_{\partial D} V^* ds.$$

Since also V^* is harmonic on $\dot{B}\setminus E$, V^* satisfies (17.1) everywhere in $\dot{B}$. So V^* is superharmonic in $\dot{B}$, as we said. Since $V^* = V$ outside of E, V^* is harmonic outside some sub-ball B_2 of $\dot{B}$. Thus Theorem 17.2 applies and yields for all $x \in \dot{B}_2$

$$V^*(x) = U^\sigma(x) + H(x),$$

where σ is a measure on $\dot{B}_2$ and H is harmonic in $\dot{B}_2$. Since V^* is harmonic in $\dot{B}\setminus E$, U^σ is also, and so σ is supported on E. Hence for $x \in \dot{B}\setminus E$

$$V(x) = U^\sigma(x) + H(x). \qquad \text{q.e.d.}$$

Note: The theorem just proved contains the result, proved in Section 11, that if V

is harmonic and bounded outside a compact set E with $C(E) = 0$, then V has a harmonic extension across E. For by the theorem, $V = U^{\sigma} + H$, σ a measure on E. Since V is bounded and H is harmonic in $\dot{B}$, U^{σ} is bounded on $\dot{B} \setminus E$, say $U^{\sigma} \leq M$.

Let $a \in E$ and suppose $U^{\sigma}(a) > M$. Then by Lemma 7.5, $U^{\sigma} > M$ in some neighborhood of a, and so at some point of $\dot{B} \setminus E$. This is false, so $U^{\sigma} \leq M$ on E. If $\sigma(E) > 0$, then $C(E) > 0$ by definition of capacity. So $\sigma = 0$, $V = H$, and we are done.

19. Wiener's Criterion

Let E be a compact subset of $\mathbb{R}^3$ and let $\bar{\mu}$ be the equilibrium distribution of unit mass on E. Then by Theorem 7.1 $\exists$ constant γ and a set $\tilde{E} \subset E$ such that $U^{\bar{\mu}} = \gamma$ on $E \setminus \tilde{E}$, $U^{\bar{\mu}} < \gamma$ on $\tilde{E}$, $C(\tilde{E}) = 0$.

Question: Given $x_0 \in E$. What is a geometric necessary and sufficient condition in order that $x_0 \in E \setminus \tilde{E}$?

Theorem 10.1 gives a sufficient condition: $x_0 \in E \setminus \tilde{E}$ provided that there is some constant $K > 0$ such that

$$\text{(1)} \qquad \frac{C(E \cap B_\rho)}{\rho} > K$$

for all small ρ, where $B_\rho = \{x \,\big|\, |x - x_0| \le \rho\}$.

We observe that (1) implies

$$\text{(2)} \qquad \int_0^1 \frac{C(E \cap B_\rho)}{\rho^2} d\rho = \infty .$$

It was discovered by Norbert Wiener that (2) is a necessary and sufficient condition for x_0 to belong to $E \setminus \tilde{E}$. In the present section, we shall prove this fact. We base our proof on the notion of thinness of a set at a point, introduced by Brelot.

Definition 19.1: Let S be a subset of $\mathbb{R}^n$, and let x be a point in the closure of S. S is thin at x if there exists a finite measure ν of compact support with $u = U^{\nu}$ such that

$$\underline{\lim}_{\substack{y \to x \\ y \in S \setminus \{x\}}} u(y) > u(x) \quad .$$

Note that $\underline{\lim}_{y \to x} u(y) = u(x)$.

Notation: Let v be a function, S a set, x a point in the closure of S . We write

$$L(v,S,x) = \underline{\lim}_{\substack{y \to x \\ y \in S \setminus \{x\}}} v(y)$$

In this notation then, S is thin at x if there exists some potential u such that

(3) $$L(u,S,x) > u(x) \quad .$$

Theorem 1: Let E *be a compact set in* $\mathbb{R}^3$ *and let* $x_0 \in E$. *Then* $x_0 \in \tilde{E}$ *if and only if* E *is thin at* x_0 .

Proof:

Assertion 1: There exists a potential u such that

(4) $u(y) = \infty$ for $y \in \tilde{E}\setminus\{x_0\}$ and $u(x_0) < \infty$.

Proof: For $n=1,2, \ldots$, put

$$K_n = \{x \in E \mid U^{\bar{\mu}}(x) \le \gamma - \frac{1}{n} , |x - x_0| \ge \frac{1}{n}\} .$$

Then K_n is compact, $C(K_n) = 0$, and $x_0 \notin K_n$. By Proposition 18.4, $\exists$ a measure μ_n of compact support with $U^{\mu_n} = \infty$ on K_n , $U^{\mu_n}(x_0) < \infty$. Without loss of generality, all the $\mu_n, n=1,2, \ldots$ have their support in a fixed ball B . Multiplying μ_n by a suitable positive constant t_n for each n , we obtain the following:

$\sum_{n=1}^{\infty} t_n \mu_n$ is a finite measure with support $\subset B$, and

$\sum_{n=1}^{\infty} t_n U^{\mu_n}(x_0) < \infty$. Put $\mu = \sum_{n=1}^{\infty} t_n \mu_n$. Fix $x \in \tilde{E}\setminus\{x_0\}$. For some n_0 , $x \in K_{n_0}$. Hence

$$U^{\mu}(x) \ge t_{n_0} U^{\mu_{n_0}}(x) = \infty .$$

So $U^{\mu} = \infty$ on $\tilde{E}\setminus\{x_0\}$, and $U^{\mu}(x_0) < \infty$. Put $u = U^{\mu}$. Then u satisfies (4), proving the Assertion.

Suppose now that $U^{\bar{\mu}}(x_0) < \gamma$. Choose $u = U^{\mu}$ satisfying (4). Put

$$\Psi = U^{\overline{\mu}+\mu} = U^{\overline{\mu}} + u \quad .$$

Then

$$L(\Psi,E,x_0) = \min[L(\Psi,\tilde{E},x_0) \ , \ L(\Psi,E\setminus\tilde{E},x_0)] \quad .$$

Since $\Psi = \infty$ on $\tilde{E}\setminus\{x_0\}$,

$$L(\Psi,E,x_0) = L(\Psi,E\setminus\tilde{E},x_0) = \gamma + L(u,E\setminus\tilde{E},x_0)$$
$$\geq \gamma + u(x_0) > U^{\overline{\mu}}(x_0) + u(x_0) = \Psi(x_0) \quad .$$

Hence E is thin at x_0. Conversely, suppose $U^{\overline{\mu}}(x_0) = \gamma$. Suppose E is thin at x_0. We must reach a contradiction. Choose $u = U^{\mu}$ with $u(x_0) < L(u,E,x_0)$. Without loss of generality, we can choose a constant M such that, putting $u' = u + M$, $u'(x_0) = -1$ and $L(u',E,x_0) > 2$. We can then choose a closed ball B centered at x_0 such that $u' \geq 1$ on $(B \cap E)\setminus\{x_0\}$. Put $v = -u'$. Then

(5) v is subharmonic in B.

(6) $v(x_0) = 1$.

(7) $v \leq -1$ on $(B \cap E)\setminus\{x_0\}$.

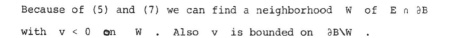

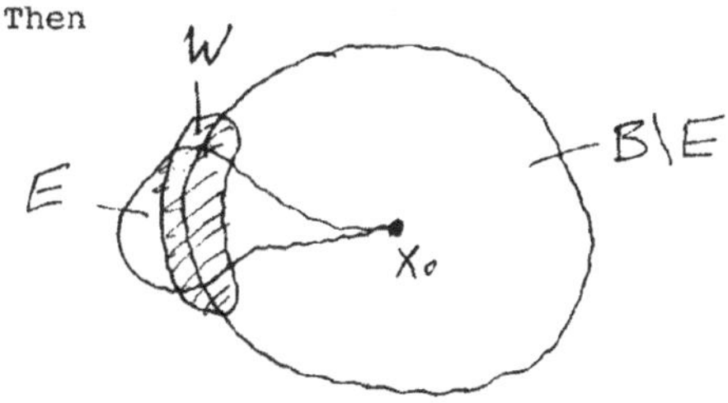

Because of (5) and (7) we can find a neighborhood W of $E \cap \partial B$ with $v < 0$ on W. Also v is bounded on $\partial B\setminus W$.

Put $w = \gamma - U^{\overline{\mu}}$. Then w is harmonic on $\mathring{B}\setminus E$, ≥ 0 on $\mathring{B}\setminus E$ and

(8) $$\lim_{y \to x_0} w(y) = 0 \quad .$$

This last equality holds since

$$\overline{\lim_{x \to x_0}}\, U^{\overline{\mu}}(x) \leq \gamma = U^{\overline{\mu}}(x_0) = \underline{\lim_{x \to x_0}}\, U^{\overline{\mu}}(x_0) \quad ,$$

and so $\lim_{x \to x_0} U^{\overline{\mu}}(x)$ exists and $= \gamma$.

w has a positive lower bound on $\partial B\setminus W$. Hence we can choose a positive constant λ such that

$$v - \lambda w \leq 0 \quad \text{on} \quad \partial B\setminus W \quad .$$

Also

$$v - \lambda w \leq 0 \quad \text{on} \quad \partial B \cap W \quad .$$

Finally, $\overline{\lim_{z \to y}}(v - \lambda w)(z) \le 0$ if $y \in \partial E \cap B$, $y \ne \{x_0\}$.

Thus $v - \lambda w$ is subharmonic on $\mathring{B} \setminus E$ and $\overline{\lim_{z \to y}}(v - \lambda w) \le 0$ for each boundary point y of $B \setminus E$, except for $y = x_0$. To take care of x_0 , we fix $\varepsilon > 0$ and we form V_ε defined for $z \in B \setminus E$ by

$$V_\varepsilon(z) = v(z) - \lambda w(z) - \frac{\varepsilon}{|z - x_0|} .$$

Then $\overline{\lim_{z \to y}} V_\varepsilon(z) \le 0$ for all $y \in \partial(\mathring{B} \setminus E)$ including $y = x_0$. By the maximum principle for subharmonic functions, then,

$$V_\varepsilon(z) \le 0 \quad , \quad z \in \mathring{B} \setminus E .$$

Letting $\varepsilon \to 0$ we conclude that

$$v(z) - \lambda w(z) \le 0 \quad , \quad z \in \mathring{B} \setminus E .$$

Letting $z \to x_0$, now, and recalling (8), we get

$$\overline{\lim_{\substack{z \to x_0 \\ z \in B \setminus E}}} v(z) \le 0$$

Also

$$\overline{\lim_{\substack{z\to x_0 \\ z\in E\setminus\{x_0\}}}} \quad v(z) \le 0 \quad .$$

Thus $\overline{\lim}_{z\to x_0} v(z) \leq 0$. But $v(x_0) = 1$. This contradicts the subharmonicity of v . Hence E is not thin at x_0 . Q.E.D.

Exercise 1: Let ϕ be an increasing function defined on $0 < r \le b$ such that $\lim_{r\to 0} \phi(r) = 0$. Then

(9) $$\int_0^b \frac{d\phi(r)}{r} = \frac{\phi(b)}{b} + \int_0^b \frac{\phi(r)}{r^2} dr \quad ,$$

the left and right sides being simultaneously finite or infinite. If the integral converges, then

(10) $$\lim_{r\to 0} \frac{\phi(r)}{r} = 0 \quad .$$

In what follows, we denote by B_ρ the ball: $|x - x_0| \le \rho$, $0 < \rho \le 1$. Further, for $r > 0$, we denote by μ_r the equilibrium distribution for $E \cap B_r$ with $U^{\mu_r} = 1$ on $E \cap B_r$ except for a set of capacity 0 .

Lemma 1: Assume that $\int_0^1 \frac{C(E \cap B_\rho)}{\rho^2} d\rho < \infty$. Then for $\varepsilon > 0$, there is $r(\varepsilon) > 0$ such that

$$U^{\mu_r}(x_0) < \varepsilon \quad \text{if} \quad r < r(\varepsilon) .$$

Proof: Fix r . Then μ_r is supported on $E \cap B_r$. Define a function ϕ on $0 \le \rho \le r$ by

$$\phi(\rho) = \mu_r(E \cap B_\rho) .$$

Then, using (9), we get

$$(11) \qquad U^{\mu_r}(x_0) = \int_0^r \frac{d\phi(\rho)}{\rho} = \frac{\phi(r)}{r} + \int_0^r \frac{\phi(\rho)}{\rho^2} d\rho .$$

For each ρ we denote by χ_ρ the characteristic function of B_ρ . Then $\chi_\rho \mu_r$ is supported on $E \cap B_\rho$ and $U^{\chi_\rho \mu_r} \le U^{\mu_r} \le 1$ on $E \cap B_\rho$. By definition of capacity, then, for $0 \le \rho \le r$,

$$(12) \qquad C(E \cap B_\rho) \ge (\chi_\rho \mu_r)(E \cap B_\rho) = \mu_r(E \cap B_\rho) = \phi(\rho) .$$

Also $\phi(r) = \mu_r(E \cap B_r) = C(E \cap B_r)$, by choice of μ_r. Hence, and using (11) and (12), we have

$$(13) \qquad U^{\mu_r}(x_0) \le \frac{C(E \cap B_r)}{r} + \int_0^r \frac{C(E \cap B_\rho)}{\rho^2} d\rho .$$

By hypothesis, $\int_0^1 \frac{C(E \cap B_\rho)}{\rho^2} d\rho < \infty$. Hence, using (10), we see that the right side in (13) tends to 0 as $r \to 0$. So Lemma 1 is proved.

<u>Lemma 2</u>: <u>Assume that</u> $\int_0^1 \frac{C(E \cap B_\rho)}{\rho^2} d\rho < \infty$. <u>Then E is thin at</u> x_0.

<u>Proof</u>: With μ_r defined as in the preceding Lemma, we choose $r > 0$ such that $U^{\mu_r}(x_0) < 1$. Furthermore, by definition of μ_r, $U^{\mu_r}(x) = 1$, $x \in E \cap B_r$, outside some set F of capacity 0. Arguing as in the proof of theorem 1, we can find a potential u with $u = \infty$ on $F \setminus \{x_0\}$ and $u(x_0) < \infty$. Put $\Psi = U^{\mu_r} + u$. To show that $L(\Psi, E, x_0) > \Psi(x_0)$ it then suffices to show that:

$$(14) \qquad L(\Psi, E \setminus F, x_0) > \Psi(x_0) .$$

$$L(\Psi, E \setminus F, x_0) = L(\Psi, (E \cap B_r) \setminus F, x_0)$$
$$= L(1 + u, (E \cap B_r) \setminus F, x_0) \ge 1 + u(x_0) > \Psi(x_0) ,$$

i.e. (14). Thus $L(\Psi,E,x_0) > \Psi(x_0)$, and so E is thin at x_0 .

Q.E.D.

<u>Lemma 3:</u> <u>Assume that</u> E <u>is thin at</u> x_0 . <u>Then there exists a potential</u> U^σ <u>with</u> $L(U^\sigma,E,x_0) = \infty$ <u>and</u> $U^\sigma(x_0) < \infty$.

<u>Proof:</u> By hypothesis, there exists a measure ν and a number $\delta > 0$, with

$$L(U^\nu - U^\nu(x_0),E,x_0) = \delta .$$

Note that $U^\nu(x_0) < \infty$.

For $j=1,2, \dots$ choose the ball

$$B_j = \{x \mid |x - x_0| \le \varepsilon_j\}$$

in such a way that $\varepsilon_j \le 1$, and

$$(15) \qquad \int_{B_j} \frac{d\nu(y)}{|y - x_0|} < \frac{1}{2^j} , \; j=1,2, \dots$$

Define by ν_j the restriction of ν to B_j , i.e. $\nu_j(S) = \nu(S \cap B_j)$, S any Borel set. Put $\sigma = \sum_{j=1}^{\infty} \nu_j$. Finally, denote B the ball of radius 1 centered at x_0 . Then σ is supported on B . In view of (15), $U^\sigma(x_0) < \infty$, and so also $\sigma(B) < \infty$.

<u>Claim:</u> $L(U^{\sigma},E,x_0) = \infty$.

For each j , we can write

$$U^{\nu} = U^{\nu_j} + h_j \quad ,$$

where h_j is continuous on $\mathring{B}_j$. Fix p .
$U^{\sigma}(y) \geq \sum_{j=1}^{p} U^{\nu_j}(y)$, for all y , and so

$$U^{\sigma}(y) - \sum_{j=1}^{p} U^{\nu_j}(x_0) \geq \sum_{j=1}^{p} (U^{\nu_j}(y) - U^{\nu_j}(x_0)) \quad .$$

Hence

$$L(U^{\sigma} - \sum_{j=1}^{p} U^{\nu_j}(x_0),E,x_0) \geq L(\sum_{j=1}^{p} (U^{\nu_j} - U^{\nu_j}(x_0)),E,x_0)$$

$$\geq \sum_{j=1}^{p} L(U^{\nu_j} - U^{\nu_j}(x_0),E,x_0) \quad .$$

For each j ,

$$L(U^{\nu_j} - U^{\nu_j}(x_0),E,x_0) = L((U^{\nu_j} + h_j) - (U^{\nu_j}(x_0) + h_j(x_0)),E,x_0)$$

$$= L(U^{\nu} - U^{\nu}(x_0),E,x_0) = \delta \quad .$$

Hence

$$L(U^{\sigma} - \sum_{j=1}^{p} U^{\nu_j}(x_0), E, x_0) \geq p\delta \quad .$$

Hence

$$L(U^{\sigma}, E, x_0) \geq p\delta \quad .$$

This holds for all p , and so $L(U^{\sigma}, E, x_0) = \infty$, as claimed.

Q.E.D.

Lemma 4: *Assume* E *is thin at* x_0 . *Then*

$$\int_0^1 \frac{C(E \cap B_t)}{t^2} dt < \infty \quad .$$

Proof: By Lemma 3, there exists a measure ν such that $U^{\nu}(x_0) < \infty$ and $L(U^{\nu}, E, x_0) = \infty$. For each $r > 0$, we denote by ν_r the restriction of ν to B_r .

Claim 1: For all sufficiently small r :

$$U^{\nu_r}(x) \geq 1 \quad , \quad x \text{ in } (B_{r/2} \cap E) \setminus (\{x_0\}) \quad .$$

We have, for all x ,

$$U^{\nu}(x) = \int_{|y-x_0|\le r} \frac{d\nu(y)}{|x - y|} + \int_{|y-x_0|>r} \frac{d\nu(y)}{|y - x|}$$

$$= U^{\nu_r}(x) + \int_{|y-x_0|>r} \frac{d\nu(y)}{|x - y|} \quad .$$

If $|x - x_0| \le r/2$ then for y in $|y - x_0| > r$, $|y - x| \ge \frac{1}{2}|y - x_0|$.

Hence

$$U^{\nu}(x) \le U^{\nu_r}(x) + 2U^{\nu}(x_0)$$

For $x \in (B_{r/2} \cap E)\setminus\{x_0\}$, then, we have, in view of $L(U^{\nu},E,x_0) = \infty$, for r small enough:

$$1 + 2U^{\nu}(x_0) < U^{\nu}(x) \le U^{\nu_r}(x) + 2U^{\nu}(x_0) \quad .$$

Hence $U^{\nu_r}(x) \ge 1$, as claimed. Choose r_0 so that $U^{\nu_{2r}} \ge 1$ on $(B_r \cap E)\setminus\{x_0\}$ for $r < r_0$.

<u>Claim 2</u>: For all $r < r_0$:

$$C(E \cap B_r) \leq \nu(B_{2r}) \quad .$$

For each r , $0 < r < r_0$, denote by μ_r the equilibrium distribution on $E \cap B_r$ with $\mu_r(E \cap B_r) = C(E \cap B_r)$. Then $U^{\mu_r} \leq 1$. We have:

$$C(E \cap B_r) = \mu_r(E \cap B_r) \leq \int_{E\cap B_r} U^{\nu_{2r}} d\mu_r \quad ,$$

since $U^{\nu_{2r}} \geq 1$ on $E \cap B_r \setminus \{x_0\}$. Hence

$$C(E \cap B_r) \leq \int_{E\cap B_r} \left\{ \int_{B_{2r}} \frac{d\nu(y)}{|x - y|} \right\} d\mu_r(x)$$

$$= \int_{B_{2r}} \left\{ \int_{E\cap B_r} \frac{d\mu_r(x)}{|x - y|} \right\} d\nu(y) \leq \int_{B_{2r}} d\nu(y) = \nu(B_{2r}) \quad ,$$

as claimed.

We put $\phi(r) = \nu(B_r)$ for all $r > 0$. Then

$$U^{\nu}(x_0) = \int_0^{\infty} \frac{d\phi(r)}{r} \quad .$$

So by choice of ν ,

(16) $$\int_0^1 \frac{d\phi(r)}{r} < \infty \quad .$$

It follows by Exercise 1 that

(17) $$\int_0^1 \frac{\phi(r)}{r^2} dr < \infty \quad .$$

Now

$$\int_0^{r_0} \frac{C(E \cap B_t)}{t^2} dt \le \int_0^{r_0} \frac{\nu(B_{2t})}{t^2} dt \quad ,$$

by Claim 2,

$$= 2\int_0^{2r_0} \frac{\nu(B_r)}{r^2} dr = 2\int_0^{2r_0} \frac{\phi(r)}{r^2} dr \quad .$$

By (17), the right hand side $< \infty$. Hence

$$\int_0^1 \frac{C(E \cap B_t)}{t^2} dt < \infty \quad .$$ Q.E.D.

We are now able to state and prove the main theorem of this section:

Wiener's Criterion: Let E be a compact set in $\mathbb{R}^3$ and let $x_0 \in E$. Then $U^{\overline{\mu}}(x_0) = \gamma$ if and only if

$$\int_0^1 \frac{C(E \cap B_\rho)}{\rho^2} d\rho = \infty \quad . \tag{18}$$

Proof: Lemmas 2 and 4 give that E is thin at x_0 if and only if

$$\int_0^1 \frac{C(E \cap B_t)}{t^2} dt < \infty \quad .$$

By Theorem 1, $U^{\overline{\mu}}(x_0) < \gamma$ if and only if E is thin at x_0 . Hence $U^{\overline{\mu}}(x_0) < \gamma$ if and only if

$$\int_0^1 \frac{C(E \cap B_t)}{t^2} dt < \infty \quad ,$$

which is equivalent to Wiener's Criterion.

We now define

$$A_n = \{x \mid \frac{1}{2^{n+1}} \le |x - x_0| \le \frac{1}{2^n}\} \text{ , } n=0,1,2, \ldots$$

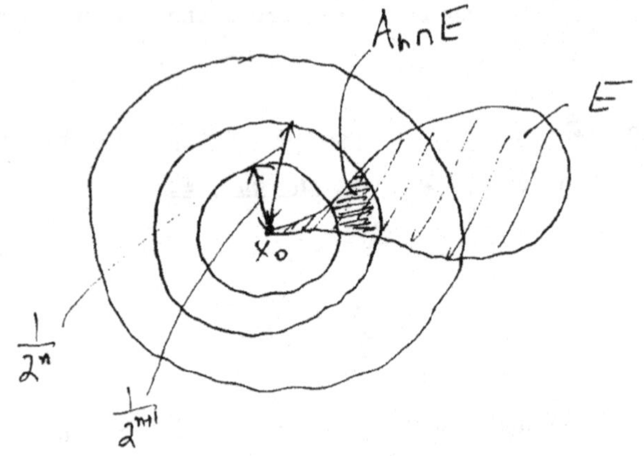

Wiener's Criterion (second version):

$U^{\overline{\mu}}(x_0) = \gamma$ if and only if

$$\sum_{n=0}^{\infty} 2^n C(A_n \cap E) = \infty \quad . \tag{19}$$

Proof: By Wiener's Criterion, $U^{\overline{\mu}}(x_0) = \gamma$ if and only if (18) holds. By Exercise 1, this is equivalent to

$$\int_0^1 \frac{dC(E \cap B_\rho)}{\rho} d\rho = \infty \quad . \tag{20}$$

$$\int_0^1 \frac{dC(E \cap B_\rho)}{\rho} = \sum_{n=0}^{\infty} \int_{\frac{1}{2^{n+1}}}^{\frac{1}{2^n}} \frac{dC(E \cap B_\rho)}{\rho}$$

$$\leq \sum_{n=0}^{\infty} 2^{n+1} \int_{\frac{1}{2^{n+1}}}^{\frac{1}{2^n}} dC(E \cap B_\rho)$$

$$= \sum_{n=0}^{\infty} 2^{n+1} \left(C(E \cap B_{\frac{1}{2^n}}) - C(E \cap B_{\frac{1}{2^{n+1}}}) \right) .$$

Now $B_{\frac{1}{2^{n+1}}} \cup A_n = B_{\frac{1}{2^n}}$.

By the subadditivity of capacity (Exercise 7.5), it follows that

$$C(E \cap B_{\frac{1}{2^n}}) - C(E \cap B_{\frac{1}{2^{n+1}}}) \leq C(E \cap A_n) .$$

Hence we get

$$\int_0^1 \frac{dC(E \cap B_\rho)}{\rho} \le \sum_{n=0}^{\infty} 2^{n+1} C(E \cap A_n) \quad .$$

If $U^{\overline{\mu}}(x_0) = \gamma$, then (20) holds and so $\sum_{n=0}^{\infty} 2^{n+1} C(E \cap A_n) = \infty$, i.e. (19).

Conversely, suppose (19) holds. For fixed n ,

$$\int_{\frac{1}{2^{n+1}}}^{2^{\frac{1}{n}}} \frac{C(E \cap B_\rho)}{\rho^2} dp \ge 2^{2n} \int_{\frac{1}{2^{n+1}}}^{2^{\frac{1}{n}}} C(E \cap B_\rho) d\rho$$

$$\ge 2^{2n}\left[\frac{1}{2^n} - \frac{1}{2^{n+1}}\right] C(E \cap B_{\frac{1}{2^{n+1}}})$$

$$\ge (2^n - 2^{n-1}) C(E \cap A_{n+1})$$

$$= 2^{n-1} C(E \cap A_{n+1}) = \tfrac{1}{4} 2^{n+1} C(E \cap A_{n+1}) \quad .$$

Hence

$$\int_0^1 \frac{C(E \cap B_\rho)}{\rho^2} d\rho = \sum_{n=0}^{\infty} \int_{\frac{1}{2^{n+1}}}^{\frac{1}{2^n}} \frac{C(E \cap B_\rho)}{\rho^2} d\rho$$

$$\geq \sum_{n=0}^{\infty} \frac{1}{4} 2^{n+1} C(E \cap A_{n+1}) = \frac{1}{4} \sum_{n=1}^{\infty} 2^n C(E \cap A_n) \quad .$$

Hence (19) implies that $\int_0^1 \frac{C(E \cap B_\rho)}{\rho^2} d\rho = \infty$, whence by Wiener's Criterion, $U^{\bar{\mu}}(x_0) = \gamma$. Thus $U^{\bar{\mu}}(x_0) = \gamma$ if and only if (19) holds. Q.E.D.

<u>Exercise 2</u>: Prove the analogue of Wiener's Criterion (19) with 2 replaced by $\lambda (\lambda > 1)$.

Appendix

We use the following notations: If $x = (x_1, \ldots, x_n) \in \mathbb{R}^n$, then

$$|x| = \left(\sum_{i=1}^{n} x_i^2\right)^{1/2} .$$

dx = Lebesgue measure on $\mathbb{R}^n$.

On $\mathbb{R}^3$ we sometimes write dV for dx.

If Σ is a smooth surface in $\mathbb{R}^n$,

dS = element of $(n-1)$-dimensional area on Σ.

If F is a set in $\mathbb{R}^n$,

$\overline{F}$ = closure of F.

$\dot{F}$ = interior of F.

∂F = boundary of F.

If X is a set,

$C(X)$ = space of all continuous functions on X.

If $\{\mu_n\}$ is a sequence of measures on X, we say "$\mu_n \to \mu$ weakly" if

$$\int f d\mu_n \to \int f d\mu \quad \text{as} \quad n \to \infty ,$$

for all $f \in C(X)$.

If Ω is a domain in $\mathbb{R}^n$,

$C_0^k(\Omega)$ = space of all k-times differentiable functions having compact support $\subset \Omega$.

$C_0^k = C_0^k(\mathbb{R}^n)$.

If μ is a measure,

supp μ = support of μ .

In what follows Ω is a smoothly bounded domain in $\mathbb{R}^n$ with $\bar{\Omega}$ compact. $\vec{N}(x)$ denotes the exterior unit normal to $\partial\Omega$ at x, and $\frac{\partial}{\partial n}$ = exterior normal derivative.

Divergence Theorem: Let $\vec{v} = (v_1, \ldots ,v_n)$ be a smooth vector-field defined on $\bar{\Omega}$. Then

$$\text{(G.1)} \qquad \int_{\partial\Omega} \vec{v}\cdot\vec{N}dS = \int_{\Omega} \operatorname{div} \vec{v}\, dx ,$$

where $\operatorname{div} \vec{v} = \sum_{i=1}^{n} \frac{\partial v_i}{\partial x_i}$.

Consequences of (G.1) are the following Green's formulae:

If u,v are smooth functions on $\bar{\Omega}$,

$$\text{(G.3)} \qquad \int_{\partial\Omega} u \frac{\partial v}{\partial n}\, dS = \int_{\Omega} \operatorname{grad} u \cdot \operatorname{grad} v\, dx + \int_{\Omega} u\Delta v dx .$$

Note: We have no (G.2).

With u,v as before

$$\text{(G.4)} \qquad \int_{\partial\Omega} \{u \frac{\partial v}{\partial n} - v \frac{\partial u}{\partial n}\} dS = \int_{\Omega} \{u\Delta v - v\Delta u\} dx$$

A function u is harmonic in Ω if $\Delta u = 0$ in Ω . If u is harmonic in Ω and B is the ball $\{x \mid |x - x_o| \leq r\} \subset \Omega$, then, with A = area of ∂B, we have:

Mean Value Property:

$$u(x_o) = \frac{1}{A} \int_{\partial B} u dS .$$

If u is harmonic in Ω and continuous in $\bar{\Omega}$, we have

Maximum Principle:

For each $x \in \Omega$,

$$\min_{\partial\Omega} u \leq u(x) \leq \max_{\partial\Omega} u .$$

Harnack's Theorem: Let $\{u_n\}$, $n = 1,2, \ldots$, be a sequence of harmonic functions in Ω which is monotone, i.e.,

$$u_1 \leq u_2 \leq \ldots \quad \text{in} \quad \Omega .$$

If $\exists x_o$ in Ω such that the sequence of numbers $\{u_n(x_o)\}$ is bounded, then $\{u_n\}$ converges uniformly on compact subsets of Ω to a harmonic limit function.

Compactness Principle: Let $\{u_n\}$, $n = 1,2, \ldots$ be a sequence of harmonic functions in Ω which is uniformly bounded, i.e., $\exists M$ with

$$|u_j(x)| \leq M, \; x \in \Omega, \quad \text{all} \quad j .$$

Then $\exists$ subsequence $\{u_{j_\nu}\}$, $\nu = 1,2, \ldots$ which converges uniformly on compact subsets of Ω to a harmonic limit function.

References

Throughout we use standard results from the calculus in several variables and from the theory of measure and integration. The book by Loomis and Sternberg, [11], could serve as a general reference for the calculus, and Royden's book, [4], as a reference for the real function theory.

Section 2

R. Becker's book, "Theorie der Elektrizität", [1], gives an excellent introduction to electrostatics. We have drawn on it, especially pp. 42-53 of [1], for Sections 2 and 3.

Section 3

See [1], p. 50.

Section 4

The theory of distributions is in L. Schwartz, [15]. Chapter II of [15] is relevant to the present Section.

Section 5

See [1], pp. 51-54 and [9], pp. 330-331.

Section 6

See Hille's book [7], Chapter 16, especially Theorem 16.4.3.

Section 7

For the main result, Theorem 7.1, see Frostman [5]. Our proof follows Carleson, [4], III.

Section 8

Theorem 8.1 is due to Evans and Vasilesco. See [6], p. 117, and p. 269. A short proof of the theorem is in [4], III. We have followed the discussion in [10], pp. 46-47.

Section 9

We have here followed [10], pp. 44-45.

Section 10

See Carleson, [4], III, Theorem 3, and Frostman, [5].

Section 11

See Carleson, [4], VII, Theorem 1.

Section 12

See Kellogg, [9], Chapter IX, Sections 3 and 4.

Section 13

See Kellogg, [9], Chapter IX, Section 2.

Section 14

Perron introduced his method of solving the boundary value problem for Laplace's equation in the paper [12].

Section 15

The notion of "barrier" is discussed in detail in [6], pp. 168-176 and a historical discussion is on p. 271. Bouligand's result (our Theorem 15.3) is given in [3].

Section 16

Kellogg's theorem is proved in [6], p. 199, Lemma 9.18. The result for two dimensions is in [8].

Section 17

Theorem 17.2 is due to F. Riesz, [13]. We have, in part, followed Helms, [6], pp. 113-115. Proposition 17.4 is due to Weyl, [16].

Section 18

Theorem 18.2 is due to Bôcher, [2].

Section 19

Wiener's Criterion (2nd version) is in N. Wiener, The Dirichlet Problem, J. Math. Phys. 3(1924), pp. 127-146. The results on thinness are due to M. Brelot, Sur les ensembles effilés, Bull. Sci. Math. France 68(1944), pp. 12-36.

Appendix

The basic properties of harmonic functions we have stated in the Section are proved in various texts, in particular see pp. 6-33 of Helms [6].

Bibliography

[1] R. Becker, "Theorie der Elektrizität", B. G. Teubner, Stuttgart, 16th edition, 1957.

[2] M. Bôcher, "Singular points of functions which satisfy partial differential equations of the elliptic type", Bull. Amer. Math. Soc., 9, 1903, pp. 455-465.

[3] G. Bouligand, "Domaines infinis et cas d'exception du problème de Dirichlet", C. R. Acad. Sci. Paris, t. 178, 1924, pp. 1054-1057.

[4] L. Carleson, "Selected Problems on Exceptional Sets", Van Nostrand Mathematical Studies #13, 1967.

[5] O. Frostman, "Potentiel d'Équilibre et Capacité des Ensembles", Dissertation, Lund, 1935.

[6] L. L. Helms, "Introduction to Potential Theory", Wiley-Interscience, 1969.

[7] E. Hille, "Analytic Function Theory", Vol. II, Ginn and Company, 1962.

[8] O. Kellogg, "Unicité des Fonctions Harmoniques", C. R. Acad. de Paris, t. 187, 1928, p. 526.

[9] O. Kellogg, "Foundations of Potential Theory", Dover Publications, 1953.+

[10] N. Ninomiya, "Potential Theory", Kyoritsu Shuppan Co., 1969 (in Japanese).

+Kellogg's book was originally published by Springer in 1929 and re-issued by Springer in 1967, in Grundlehren der mathematischen Wissenschaften. Our page references are to the Dover edition, 1953.

[11] L. Loomis and S. Sternberg, "Advanced Calculus", Addison-Wesley, 1968.

[12] O. Perron, "Eine neue Behandlung der ersten Randwertaufgabe für $\Delta u = 0$", Math. Zeitschrift, vol. 18, 1923, pp. 42-54.

[13] F. Riesz, "Sur les fonctions subharmoniques et leur rapport à la théorie du potentiel", Acta Math., vol. 54, 1930, pp. 321-360.

[14] H. Royden, "Real Analysis", The Macmillan Co., 1963.

[15] L. Schwartz, "Théorie des Distributions", tome 1, Hermann, 1957.

[16] H. Weyl, "Method of orthogonal projections in potential theory", Duke Math. J., 7, 1940, pp. 411-444.

Index

Vol. 670: Fonctions de Plusieurs Variables Complexes III, Proceedings, 1977. Edité par F. Norguet. XII, 394 pages. 1978.

Vol. 671: R. T. Smythe and J. C. Wierman, First-Passage Perculation on the Square Lattice. VIII, 196 pages. 1978.

Vol. 672: R. L. Taylor, Stochastic Convergence of Weighted Sums of Random Elements in Linear Spaces. VII, 216 pages. 1978.

Vol. 673: Algebraic Topology, Proceedings 1977. Edited by P. Hoffman, R. Piccinini and D. Sjerve. VI, 278 pages. 1978.

Vol. 674: Z. Fiedorowicz and S. Priddy, Homology of Classical Groups Over Finite Fields and Their Associated Infinite Loop Spaces. VI, 434 pages. 1978.

Vol. 675: J. Galambos and S. Kotz, Characterizations of Probability Distributions. VIII, 169 pages. 1978.

Vol. 676: Differential Geometrical Methods in Mathematical Physics II, Proceedings, 1977. Edited by K. Bleuler, H. R. Petry and A. Reetz. VI, 626 pages. 1978.

Vol. 677: Séminaire Bourbaki, vol. 1976/77, Exposés 489–506. IV, 264 pages. 1978.

Vol. 678: D. Dacunha-Castelle, H. Heyer et B. Roynette. Ecole d'Ete de Probabilités de Saint-Flour. VII-1977. Edité par P. L. Hennequin. IX, 379 pages. 1978.

Vol. 679: Numerical Treatment of Differential Equations in Applications, Proceedings, 1977. Edited by R. Ansorge and W. Törnig. IX, 163 pages. 1978.

Vol. 680: Mathematical Control Theory, Proceedings, 1977. Edited by W. A. Coppel. IX, 257 pages. 1978.

Vol. 681: Séminaire de Théorie du Potentiel Paris, No. 3, Directeurs: M. Brelot, G. Choquet et J. Deny. Rédacteurs: F. Hirsch et G. Mokobodzki. VII, 294 pages. 1978.

Vol. 682: G. D. James, The Representation Theory of the Symmetric Groups. V, 156 pages. 1978.

Vol. 683: Variétés Analytiques Compactes, Proceedings, 1977. Edité par Y. Hervier et A. Hirschowitz. V, 248 pages. 1978.

Vol. 684: E. E. Rosinger, Distributions and Nonlinear Partial Differential Equations. XI, 146 pages. 1978.

Vol. 685: Knot Theory, Proceedings, 1977. Edited by J. C. Hausmann. VII, 311 pages. 1978.

Vol. 686: Combinatorial Mathematics, Proceedings, 1977. Edited by D. A. Holton and J. Seberry. IX, 353 pages. 1978.

Vol. 687: Algebraic Geometry, Proceedings, 1977. Edited by L. D. Olson. V, 244 pages. 1978.

Vol. 688: J. Dydak and J. Segal, Shape Theory. VI, 150 pages. 1978.

Vol. 689: Cabal Seminar 76–77, Proceedings, 1976–77. Edited by A.S. Kechris and Y. N. Moschovakis. V, 282 pages. 1978.

Vol. 690: W. J. J. Rey, Robust Statistical Methods. VI, 128 pages. 1978.

Vol. 691: G. Viennot, Algèbres de Lie Libres et Monoïdes Libres. III, 124 pages. 1978.

Vol. 692: T. Husain and S. M. Khaleelulla, Barrelledness in Topological and Ordered Vector Spaces. IX, 258 pages. 1978.

Vol. 693: Hilbert Space Operators, Proceedings, 1977. Edited by J. M. Bachar Jr. and D. W. Hadwin. VIII, 184 pages. 1978.

Vol. 694: Séminaire Pierre Lelong – Henri Skoda (Analyse) Année 1976/77. VII, 334 pages. 1978.

Vol. 695: Measure Theory Applications to Stochastic Analysis, Proceedings, 1977. Edited by G. Kallianpur and D. Kölzow. XII, 261 pages. 1978.

Vol. 696: P. J. Feinsilver, Special Functions, Probability Semigroups, and Hamiltonian Flows. VI, 112 pages. 1978.

Vol. 697: Topics in Algebra, Proceedings, 1978. Edited by M. F. Newman. XI, 229 pages. 1978.

Vol. 698: E. Grosswald, Bessel Polynomials. XIV, 182 pages. 1978.

Vol. 699: R. E. Greene and H.-H. Wu, Function Theory on Manifolds Which Possess a Pole. III, 215 pages. 1979.

Vol. 700: Module Theory, Proceedings, 1977. Edited by C. Faith and S. Wiegand. X, 239 pages. 1979.

Vol. 701: Functional Analysis Methods in Numerical Analysis, Proceedings, 1977. Edited by M. Zuhair Nashed. VII, 333 pages. 1979.

Vol. 702: Yuri N. Bibikov, Local Theory of Nonlinear Analytic Ordinary Differential Equations. IX, 147 pages. 1979.

Vol. 703: Equadiff IV, Proceedings, 1977. Edited by J. Fábera. XIX, 441 pages. 1979.

Vol. 704: Computing Methods in Applied Sciences and Engineering, 1977, I. Proceedings, 1977. Edited by R. Glowinski and J. L. Lions. VI, 391 pages. 1979.

Vol. 705: O. Forster und K. Knorr, Konstruktion verseller Familien kompakter komplexer Räume. VII, 141 Seiten. 1979.

Vol. 706: Probability Measures on Groups, Proceedings, 1978. Edited by H. Heyer. XIII, 348 pages. 1979.

Vol. 707: R. Zielke, Discontinuous Čebyšev Systems. VI, 111 pages. 1979.

Vol. 708: J. P. Jouanolou, Equations de Pfaff algébriques. V, 255 pages. 1979.

Vol. 709: Probability in Banach Spaces II. Proceedings, 1978. Edited by A. Beck. V, 205 pages. 1979.

Vol. 710: Séminaire Bourbaki vol. 1977/78, Exposés 507–524. IV, 328 pages. 1979.

Vol. 711: Asymptotic Analysis. Edited by F. Verhulst. V, 240 pages. 1979.

Vol. 712: Equations Différentielles et Systèmes de Pfaff dans le Champ Complexe. Edité par R. Gérard et J.-P. Ramis. V, 364 pages. 1979.

Vol. 713: Séminaire de Théorie du Potentiel, Paris No. 4. Edité par F. Hirsch et G. Mokobodzki. VII, 281 pages. 1979.

Vol. 714: J. Jacod, Calcul Stochastique et Problèmes de Martingales. X, 539 pages. 1979.

Vol. 715: Inder Bir S. Passi, Group Rings and Their Augmentation Ideals. VI, 137 pages. 1979.

Vol. 716: M. A. Scheunert, The Theory of Lie Superalgebras. X, 271 pages. 1979.

Vol. 717: Grosser, Bidualräume und Vervollständigungen von Banachmoduln. III, 209 pages. 1979.

Vol. 718: J. Ferrante and C. W. Rackoff, The Computational Complexity of Logical Theories. X, 243 pages. 1979.

Vol. 719: Categorial Topology, Proceedings, 1978. Edited by H. Herrlich and G. Preuß. XII, 420 pages. 1979.

Vol. 720: E. Dubinsky, The Structure of Nuclear Fréchet Spaces. V, 187 pages. 1979.

Vol. 721: Séminaire de Probabilités XIII. Proceedings, Strasbourg, 1977/78. Edité par C. Dellacherie, P. A. Meyer et M. Weil. VII, 647 pages. 1979.

Vol. 722: Topology of Low-Dimensional Manifolds. Proceedings, 1977. Edited by R. Fenn. VI, 154 pages. 1979.

Vol. 723: W. Brandal, Commutative Rings whose Finitely Generated Modules Decompose. II, 116 pages. 1979.

Vol. 724: D. Griffeath, Additive and Cancellative Interacting Particle Systems. V, 108 pages. 1979.

Vol. 725: Algèbres d'Opérateurs. Proceedings, 1978. Edité par P. de la Harpe. VII, 309 pages. 1979.

Vol. 726: Y.-C. Wong, Schwartz Spaces, Nuclear Spaces and Tensor Products. VI, 418 pages. 1979.

Vol. 727: Y. Saito, Spectral Representations for Schrödinger Operators With Long-Range Potentials. V, 149 pages. 1979.

Vol. 728: Non-Commutative Harmonic Analysis. Proceedings, 1978. Edited by J. Carmona and M. Vergne. V, 244 pages. 1979.

Vol. 729: Ergodic Theory. Proceedings, 1978. Edited by M. Denker and K. Jacobs. XII, 209 pages. 1979.

Vol. 730: Functional Differential Equations and Approximation of Fixed Points. Proceedings, 1978. Edited by H.-O. Peitgen and H.-O. Walther. XV, 503 pages. 1979.

Vol. 731: Y. Nakagami and M. Takesaki, Duality for Crossed Products of von Neumann Algebras. IX, 139 pages. 1979.

Vol. 732: Algebraic Geometry. Proceedings, 1978. Edited by K. Lønsted. IV, 658 pages. 1979.

Vol. 733: F. Bloom, Modern Differential Geometric Techniques in the Theory of Continuous Distributions of Dislocations. XII, 206 pages. 1979.

Vol. 734: Ring Theory, Waterloo, 1978. Proceedings, 1978. Edited by D. Handelman and J. Lawrence. XI, 352 pages. 1979.

Vol. 735: B. Aupetit, Propriétés Spectrales des Algèbres de Banach. XII, 192 pages. 1979.

Vol. 736: E. Behrends, M-Structure and the Banach-Stone Theorem. X, 217 pages. 1979.

Vol. 737: Volterra Equations. Proceedings 1978. Edited by S.-O. Londen and O. J. Staffans. VIII, 314 pages. 1979.

Vol. 738: P. E. Conner, Differentiable Periodic Maps. 2nd edition, IV, 181 pages. 1979.

Vol. 739: Analyse Harmonique sur les Groupes de Lie II. Proceedings, 1976-78. Edited by P. Eymard et al. VI, 646 pages. 1979.

Vol. 740: Séminaire d'Algèbre Paul Dubreil. Proceedings, 1977-78. Edited by M.-P. Malliavin. V, 456 pages. 1979.

Vol. 741: Algebraic Topology, Waterloo 1978. Proceedings. Edited by P. Hoffman and V. Snaith. XI, 655 pages. 1979.

Vol. 742: K. Clancey, Seminormal Operators. VII, 125 pages. 1979.

Vol. 743: Romanian-Finnish Seminar on Complex Analysis. Proceedings, 1976. Edited by C. Andreian Cazacu et al. XVI, 713 pages. 1979.

Vol. 744: I. Reiner and K. W. Roggenkamp, Integral Representations. VIII, 275 pages. 1979.

Vol. 745: D. K. Haley, Equational Compactness in Rings. III, 167 pages. 1979.

Vol. 746: P. Hoffman, τ-Rings and Wreath Product Representations. V, 148 pages. 1979.

Vol. 747: Complex Analysis, Joensuu 1978. Proceedings, 1978. Edited by I. Laine, O. Lehto and T. Sorvali. XV, 450 pages. 1979.

Vol. 748: Combinatorial Mathematics VI. Proceedings, 1978. Edited by A. F. Horadam and W. D. Wallis. IX, 206 pages. 1979.

Vol. 749: V. Girault and P.-A. Raviart, Finite Element Approximation of the Navier-Stokes Equations. VII, 200 pages. 1979.

Vol. 750: J. C. Jantzen, Moduln mit einem höchsten Gewicht. III, 195 Seiten. 1979.

Vol. 751: Number Theory, Carbondale 1979. Proceedings. Edited by M. B. Nathanson. V, 342 pages. 1979.

Vol. 752: M. Barr, *-Autonomous Categories. VI, 140 pages. 1979.

Vol. 753: Applications of Sheaves. Proceedings, 1977. Edited by M. Fourman, C. Mulvey and D. Scott. XIV, 779 pages. 1979.

Vol. 754: O. A. Laudal, Formal Moduli of Algebraic Structures. III, 161 pages. 1979.

Vol. 755: Global Analysis. Proceedings, 1978. Edited by M. Grmela and J. E. Marsden. VII, 377 pages. 1979.

Vol. 756: H. O. Cordes, Elliptic Pseudo-Differential Operators – An Abstract Theory. IX, 331 pages. 1979.

Vol. 757: Smoothing Techniques for Curve Estimation. Proceedings, 1979. Edited by Th. Gasser and M. Rosenblatt. V, 245 pages. 1979.

Vol. 758: C. Năstăsescu and F. Van Oystaeyen; Graded and Filtered Rings and Modules. X, 148 pages. 1979.

Vol. 759: R. L. Epstein, Degrees of Unsolvability: Structure and Theory. XIV, 216 pages. 1979.

Vol. 760: H.-O. Georgii, Canonical Gibbs Measures. VIII, 190 pages. 1979.

Vol. 761: K. Johannson, Homotopy Equivalences of 3-Manifolds with Boundaries. 2, 303 pages. 1979.

Vol. 762: D. H. Sattinger, Group Theoretic Methods in Bifurcation Theory. V, 241 pages. 1979.

Vol. 763: Algebraic Topology, Aarhus 1978. Proceedings, 1978. Edited by J. L. Dupont and H. Madsen. VI, 695 pages. 1979.

Vol. 764: B. Srinivasan, Representations of Finite Chevalley Groups. XI, 177 pages. 1979.

Vol. 765: Padé Approximation and its Applications. Proceedings, 1979. Edited by L. Wuytack. VI, 392 pages. 1979.

Vol. 766: T. tom Dieck, Transformation Groups and Representation Theory. VIII, 309 pages. 1979.

Vol. 767: M. Namba, Families of Meromorphic Functions on Compact Riemann Surfaces. XII, 284 pages. 1979.

Vol. 768: R. S. Doran and J. Wichmann, Approximate Identities and Factorization in Banach Modules. X, 305 pages. 1979.

Vol. 769: J. Flum, M. Ziegler, Topological Model Theory. X, 151 pages. 1980.

Vol. 770: Séminaire Bourbaki vol. 1978/79 Exposés 525-542. IV, 341 pages. 1980.

Vol. 771: Approximation Methods for Navier-Stokes Problems. Proceedings, 1979. Edited by R. Rautmann. XVI, 581 pages. 1980.

Vol. 772: J. P. Levine, Algebraic Structure of Knot Modules. XI, 104 pages. 1980.

Vol. 773: Numerical Analysis. Proceedings, 1979. Edited by G. A. Watson. X, 184 pages. 1980.

Vol. 774: R. Azencott, Y. Guivarc'h, R. F. Gundy, Ecole d'Eté de Probabilités de Saint-Flour VIII-1978. Edited by P. L. Hennequin. XIII, 334 pages. 1980.

Vol. 775: Geometric Methods in Mathematical Physics. Proceedings, 1979. Edited by G. Kaiser and J. E. Marsden. VII, 257 pages. 1980.

Vol. 776: B. Gross, Arithmetic on Elliptic Curves with Complex Multiplication. V, 95 pages. 1980.

Vol. 777: Séminaire sur les Singularités des Surfaces. Proceedings, 1976-1977. Edited by M. Demazure, H. Pinkham and B. Teissier. IX, 339 pages. 1980.

Vol. 778: SK1 von Schiefkörpern. Proceedings, 1976. Edited by P. Draxl and M. Kneser. II, 124 pages. 1980.

Vol. 779: Euclidean Harmonic Analysis. Proceedings, 1979. Edited by J. J. Benedetto. III, 177 pages. 1980.

Vol. 780: L. Schwartz, Semi-Martingales sur des Variétés, et Martingales Conformes sur des Variétés Analytiques Complexes. XV, 132 pages. 1980.

Vol. 781: Harmonic Analysis Iraklion 1978. Proceedings 1978. Edited by N. Petridis, S. K. Pichorides and N. Varopoulos. V, 213 pages. 1980.

Vol. 782: Bifurcation and Nonlinear Eigenvalue Problems. Proceedings, 1978. Edited by C. Bardos, J. M. Lasry and M. Schatzman. VIII, 296 pages. 1980.

Vol. 783: A. Dinghas, Wertverteilung meromorpher Funktionen in ein- und mehrfach zusammenhängenden Gebieten. Edited by R. Nevanlinna and C. Andreian Cazacu. XIII, 145 pages. 1980.

Vol. 784: Séminaire de Probabilités XIV. Proceedings, 1978/79. Edited by J. Azéma and M. Yor. VIII, 546 pages. 1980.

Vol. 785: W. M. Schmidt, Diophantine Approximation. X, 299 pages. 1980.

Vol. 786: I. J. Maddox, Infinite Matrices of Operators. V, 122 pages. 1980.